Learning Geography Beyond the Traditional Classroom

Chew-Hung Chang · Bing Sheng Wu
Tricia Seow · Kim Irvine
Editors

Learning Geography Beyond the Traditional Classroom

Examples from Peninsular Southeast Asia

Editors
Chew-Hung Chang
National Institute of Education
Nanyang Technological University
Singapore
Singapore

Bing Sheng Wu
National Taiwan Normal University
Taipei
Taiwan

Tricia Seow
National Institute of Education
Nanyang Technological University
Singapore
Singapore

Kim Irvine
National Institute of Education
Nanyang Technological University
Singapore
Singapore

ISBN 978-981-10-8704-2   ISBN 978-981-10-8705-9 (eBook)
https://doi.org/10.1007/978-981-10-8705-9

Library of Congress Control Number: 2018934434

© Springer Nature Singapore Pte Ltd. 2018
This work is subject to copyright. All rights are reserved by the Publisher, whether the whole or part of the material is concerned, specifically the rights of translation, reprinting, reuse of illustrations, recitation, broadcasting, reproduction on microfilms or in any other physical way, and transmission or information storage and retrieval, electronic adaptation, computer software, or by similar or dissimilar methodology now known or hereafter developed.
The use of general descriptive names, registered names, trademarks, service marks, etc. in this publication does not imply, even in the absence of a specific statement, that such names are exempt from the relevant protective laws and regulations and therefore free for general use.
The publisher, the authors and the editors are safe to assume that the advice and information in this book are believed to be true and accurate at the date of publication. Neither the publisher nor the authors or the editors give a warranty, express or implied, with respect to the material contained herein or for any errors or omissions that may have been made. The publisher remains neutral with regard to jurisdictional claims in published maps and institutional affiliations.

Printed on acid-free paper

This Springer imprint is published by the registered company Springer Nature Singapore Pte Ltd. part of Springer Nature
The registered company address is: 152 Beach Road, #21-01/04 Gateway East, Singapore 189721, Singapore

Preface

This book brings together editors from across physical and human geography as well as geography education. The original intention of the team was to put together an edited volume of works from Southeast Asia, arising out of a series of biennial conferences organized by the Southeast Asian Geography Association. As with all edited book volumes, the final lineup of chapters was a result of variable response to the call for papers, a rigorous review process and the prudent selection of chapters that address the theme. While the editorial team was very keen to include as many examples from the region as possible, we were ultimately limited by the range and scope of topics that would fit the theme of the book.

The book is titled "Learning Geography Beyond the Traditional Classroom—Examples from Peninsular Southeast Asia", with a focus on fieldwork and use of technology in geography from examples in Singapore, Malaysia and Thailand. While Peninsular Southeast Asia encompasses a region wider than the countries mentioned, the countries mentioned also extend beyond the mainland to Archipelago Southeast Asia. This notion is probably fodder for much geographical debate, but the scope of the book is to present a range of examples about how geography can be taught and learnt well beyond the classroom, at least in the countries selected. Four questions come to mind when we ask how geography can be taught better:

What to teach?—What do our students need to learn about the geography subject?
Who we teach?—Who are our learners and how do they learn geography best?
How do we teach?—What are the strategies/techniques we can use to teach geography for deeper understanding?
Where do we teach it?—How can geography be learnt outside the traditional classroom?
How do we know that the learners have learnt?—What are the evidence of learning/assessment?

The "What" and "Who" parts of the book address the relationships between two fundamental aspects of teaching and learning: the subject matter knowledge of the geography curriculum and the needs of diverse learners. The "How" part focuses on learning design, and its emphasis is on designing thoughtful learning activities that will promote student understanding in the classroom. The "How do we know" focuses on the various forms of formative and summative assessment to check for students' understanding. Consequently, the book is divided into four main sections that address the theoretical aspects of teaching and learning geography with fieldwork and technology, examples of learning geography through fieldwork and examples of learning with information and communication technology. These sections are constructed to provide for ease of reading, and one must be mindful that there are clear areas of overlap across the chapters.

The key goal of the book is to provide examples that will help educators and education researchers to reflect on their work in advancing geography education. While the examples provided are by no means exhaustive, the editors would urge the reader to extend the ideas and issues presented into their own domains of work, whether that of a teacher, a teacher leader, a teacher educator, an education researcher or an academic geographer. Given the unprecedented changes to our global physical, social, political and economic environments, and the variable impact to communities at the national and even local scales, geographical education will empower our future generations with the knowledge, skills and attitudes to engage these uncertain and complex issues. We have implicitly accepted the importance of fieldwork and argued that using technology for learning geography will happen, just because technology is already available. At the same time, just as the carpenter must select the right tool for the job, the educator also must be mindful that the application of a particular technology must make sense and not simply become technology for technology's sake. It is timely that we should reexamine these notions and discuss how these practices can be used efficaciously to advance geographical education.

Singapore, Singapore	Chew-Hung Chang
Taipei, Taiwan	Bing Sheng Wu
Singapore, Singapore	Tricia Seow
Singapore, Singapore	Kim Irvine

Contents

Part I What Do We Know About Teaching and Learning Geography Beyond the Traditional Classroom?

1. **The Where and How of Learning Geography Beyond the Classroom** 3
   Chew-Hung Chang, Bing Sheng Wu, Tricia Seow and Kim Irvine

2. **Learning in the Field—A Conceptual Approach to Field-Based Learning in Geography** 11
   Diganta Das and Kalyani Chatterjea

3. **Teaching Geography with Technology—A Critical Commentary** 35
   Chew-Hung Chang and Bing Sheng Wu

4. **The Importance of Assessing How Geography Is Learnt Beyond the Classroom** 49
   Chew-Hung Chang and Tricia Seow

Part II Teaching and Learning Geography Through Field Inquiry

5. **What Happened to the Textbook Example of the Padang Benggali Groyne Field in Butterworth, Penang?** 63
   Tiong Sa Teh

6. **The River Guardian Program for Junior High Schools on the "River of Kings," Thailand** 79
   Supitcha Kiatprajak and Lynda Rolph

7. **Paradigm Shift in Humanities Learning Journey** 101
   Marc Teng and Kah Chee Chan

Part III Teaching and Learning Geography with Information and Communication Technology

8 Authentic Learning: Making Sense of the Real Environment Using Mobile Technology Tool 111
Kalyani Chatterjea

9 Location-Aware, Context-Rich Field Data Recording, Using Mobile Devices for Field-Based Learning in Geography 133
Muhammad Faisal Bin Aman and Boon Kiat Tay

10 High-Speed Mobile Telecommunication Technology in the Geography Classroom 145
Shahul Hameed, Pillai Vidhu and Tan Xue Ling Sherlyn Theresia

Part IV Reflecting on Factors that Support Teaching and Learning Geography Beyond the Traditional Classroom

11 Social Media—A Space to Learn 163
Frances Ess

12 Teachers' Comfort Level and School Support in Using Information and Communications Technology (ICT) to Enhance Spatial Thinking 173
Zhang Hua'an Noah and Tricia Seow

13 Reflecting on Field-Based and Technology-Enabled Learning in Geography 201
Chew-Hung Chang, Kim Irvine, Bing Sheng Wu and Tricia Seow

Index 213

Editors and Contributors

About the Editors

Chew-Hung Chang is a geography educator serving as the Co-Chair of the International Geographical Union, Commission on Geographical Education, Co-Editor of the journal International Research in Geographical and Environmental Education, as well as the President of the Southeast Asian Geography Association. In addition to being a teacher educator, he has published extensively across areas in geography, climate change education, environmental and sustainability education.

Bing Sheng Wu has an expertise in Geographic Information Science and Geo-Spatial Technologies. He teaches across undergraduate to graduate level courses at the National Taiwan Normal University and has been actively researching on applied aspects of Geo-Spatial Technologies for humanities education.

Tricia Seow is an experienced geography educator, and she is involved in the Fieldwork Exercise Task Force of the International Geography Olympiad and in the MOE Humanities Talent Development Programme. She is also the Hon-Gen Secretary of the Southeast Asian Geography Association. She enjoys fieldwork, and her research interest includes teachers' knowledge and practice, field inquiry in geography and climate change education.

Kim Irvine has worked in the field of hydrology and water resources for more than 30 years. He was awarded the New York Water Environment Association, Environmental Science Award, in 2013. His work has ranged from detailed regulatory studies to capacity building workshops on water quality sampling and assessment for universities, NGOs and government agencies throughout North America and Southeast Asia. His research interests include urban hydrology, water resource management, water quality and applied urban drainage modelling.

Contributors

Muhammad Faisal Bin Aman National Institute of Education, Nanyang Technological University, Singapore, Singapore

Kah Chee Chan Wholistic Learning Consortium, Singapore, Singapore

Chew-Hung Chang National Institute of Education, Nanyang Technological University, Singapore, Singapore

Kalyani Chatterjea National Institute of Education, Nanyang Technological University, Singapore, Singapore

Diganta Das National Institute of Education, Nanyang Technological University, Singapore, Singapore

Frances Ess Mayflower Secondary School, Singapore, Singapore

Shahul Hameed Pierce Secondary School, Singapore, Singapore

Kim Irvine National Institute of Education, Nanyang Technological University, Singapore, Singapore

Supitcha Kiatprajak Traidhos Three-Generation Community for Learning, Chiang Mai, Thailand

Zhang Hua'an Noah National Institute of Education, Nanyang Technological University, Singapore, Singapore

Lynda Rolph Traidhos Three-Generation Community for Learning, Chiang Mai, Thailand

Tricia Seow National Institute of Education, Nanyang Technological University, Singapore, Singapore

Tan Xue Ling Sherlyn Theresia Pierce Secondary School, Singapore, Singapore

Boon Kiat Tay Marsiling Secondary School, Singapore, Singapore

Tiong Sa Teh Singapore, Singapore

Marc Teng STA Travel, Singapore, Singapore

Pillai Vidhu Pierce Secondary School, Singapore, Singapore

Bing Sheng Wu National Taiwan Normal University, Taipei, Taiwan

Part I
What Do We Know About Teaching and Learning Geography Beyond the Traditional Classroom?

As the first part of the book, these four chapters provide the theoretical and empirical background for what teaching and learning geography beyond the traditional classroom means. The two modalities of learning geography with fieldwork and technology "beyond the traditional classroom" are discussed in Chaps. 2 and 3, and the issue of assessment of geography learning in these contexts is discussed in Chap. 4. Chapter 1 outlines the evolution of school geography curricula and the increased focus on topics such as environmental change and globalisation. Education needs to go beyond learning how to read and write; it needs to change the child from a state of daily routine and encounters to one where disciplinary knowledge empowers the child to imagine his or her own future. This has created a rising demand to prepare teachers to conduct fieldwork and to design technology-enabled lessons.

Chapter 2 provides a conceptualization of how an effective field learning experience (in both human and physical geographies) can be conducted. This chapter provides a clear framework for teachers to conduct meaningful learning experiences of geography in the field; to develop a question, to gather and collect data, to process and reorganize the data, and to reflect and make sense of the information collected. There are examples on how students get trained in different ways of conducting fieldwork, understanding the field setting, learning how to operate equipment (for physical geography fieldwork and various other fields). In addition, fieldwork not only develops students' critical thinking and motivates them towards better conceptual engagement, it also provides opportunities for learning better, to learn to manage time and resources, to learn leadership skills, to negotiate different geographies, to appreciate culture and to make long-lasting friendships.

Chapter 3 discusses how ICT can be used for learning in geography and how to encourage students to learn beyond facts and analyse and apply what they have learnt. Technology should not inhibit learning but should be harnessed to provide the student with the greatest potential to learn geography. While technology has expanded the range of information sources and resources for the teacher, there should not be wholesale adoption without customization. This chapter continues to discuss the role of teachers as curriculum makers and the importance of choosing the right technology, the learning activity, the key concepts and how they can help students think geographically to take them beyond what they already know.

Chapter 4 provides the key dimensions to evaluate what good geographical assessment entails, and then extending these to assessment in the field and in using ICT. It discusses the challenge of determining what good assessment is and how to extend it beyond the classroom, whether good assessment practices in geography will allow the teacher to determine how well they are teaching and how well the students are learning. In addition, assessment should be considered as an integral part of the curriculum making process, because it helps in the design of instruction that is aligned to the intended outcomes, from the cognitive and skills domains to behavioural and attitudinal outcomes, of the geography curriculum.

Chapter 1
The Where and How of Learning Geography Beyond the Classroom

Chew-Hung Chang, Bing Sheng Wu, Tricia Seow and Kim Irvine

Abstract School geography curricula have been evolving in keeping abreast with the issues that affect humankind in a fast-changing world. Key elements of this evolution include increased focus on topics such as environmental change and globalisation. Furthermore, there is a more explicit articulation on the modes of instruction in curricula documents, often expounding the virtues of technology and field-based learning. This has resulted in a proliferation of ideas in response to a rising demand to prepare teachers to conduct fieldwork and to design technology-enabled lessons. What is intuitive but often ignored is that while the context of learning has been transplanted from the traditional classrooms into new spaces—the field and cyberspace—that the teaching and learning of geography exist within the framework of formal curriculum, in as far as teacher taught activities are concerned. This book provides a collection of critical pieces that support the idea that good teaching and learning of geography in fieldwork and using technology should consider the dimensions of curriculum design, instructional design and resource provision, as well as assessment for such learning activities. Further, the book is organised to clearly describe the thinking, experiences and critical comments to two broad areas of learning outside the traditional classroom—the field and technology.

C.-H. Chang (✉) · T. Seow (✉) · K. Irvine (✉)
National Institute of Education, Nanyang Technological University,
Singapore, Singapore
e-mail: chewhung.chang@nie.edu.sg

T. Seow
e-mail: tricia.seow@nie.edu.sg

K. Irvine
e-mail: kim.irvine@nie.edu.sg

B. S. Wu (✉)
National Taiwan Normal University, Taipei, Taiwan
e-mail: wbs@ntnu.edu.tw

Why Does Geography Education Matter?

Geography is a future-oriented discipline. Over the past three decades, geography educators have been concerned with developing school geography that can enable and empower students for the future. In the 1990s, a decade where awareness of environmental issues was gaining momentum after watershed publications that included "Global Warming—the Greenpeace Report" (Leggett 1990) and "Report of the World Commission on Environment and Development: our common future" (Brundtland 1987), the focus of geographical education was to help students develop "their'geographic antennae' and to bring geographic dimensions of all their activities and all events around them into conscious awareness" (Romey and Elberty Jr 1989). At the turn of the millennium, the concern shifted to the need to keep up with "an ever burgeoning … research frontier in academic geography" (Kent 2000). A range of issues for geographical education was highlighted by Gerber (2001) including the relative importance of knowledge, skills and values, curriculum framing, as well as the use of various teaching methods such as fieldwork and technology. Indeed, "[t]he main purpose of education is to help (young) people to be prepared for today and tomorrow" (Béneker and van der Schee 2015, p. 287). Certainly, the perspectives presented here are but a sample of the range of issues discussed by the Southeast Asian community of geography educators. However, an apparent theme that runs through these past few decades is that school geography is concerned with empowering learners with the knowledge, skills and values to engage the issues of our time, through various authentic and relevant methods of teaching and learning.

The International Charter on Geographical Education affirms that "geographical education is indispensable to the development of responsible and active citizens in the present and future world" (International Geographical Union—Commission on Geographical Education 2016). Education's ultimate goal is to empower the learner to become a responsible and active citizen as they function and succeed in society. This is predicated on understanding the knowledge, skills and values that the learner needs. Education sociologist Michael Young proposes an interesting relationship between knowledge, curriculum and the future school or a "Three Futures" approach to curriculum. This approach was based on a review of the relationship between knowledge taught in school and the child's everyday experiences and three futures scenarios (Muller and Young 2008). This approach focuses on the relationship between knowledge and the curriculum and is described broadly by the school and non-school boundaries of school knowledge in three models, namely

Future 1: boundaries as given,
Future 2: a boundary-less world and
Future 3: the idea of boundary maintenance as a condition for boundary crossing

In the Future 1 model, knowledge is inherited and there is little room for change or development. Indeed, the transmission of knowledge is crucial to maintaining the boundaries of school knowledge in the F1 model. In the Future 2 model, the goals of education are to ensure the employability of students when they graduate from the schools. With increased focus on applied subjects and vocational training, the school subject disciplines are weakened. A typical response to the importance of learning subject discipline knowledge might be that there is no need to learn facts as one can always "google it". In an F2 curriculum, learners do not see a need for disciplinary knowledge and can get by with vocational training and everyday knowledge. The problem with this model is that people do not know what they do not know. There is a danger of an over-socialisation of knowledge, uncritical acceptance of unreliable sources or cynicism over information found on the Internet. This debate has been brought to the fore during the 2017 US presidential election. The F2 future may encourage the learner to become overly sceptical, criticise and refute all knowledge that is taught in schools, very likely referring to their own constructed naive theories. A middle ground is needed to help learners engage the information they encounter, within the contextual understanding of school knowledge, and ask critical questions that will develop deeper understanding of the issue at hand. Such is the Future 3 or F3 curriculum proposed by Michael Young (Young 2014). Unlike Future 1, the boundaries of knowledge do not stay stagnant in F3. Disciplinary subjects are supported and challenged by the "discoveries by members of the disciplinary communities [academics], that are associated with and by the research undertaken by associations of teachers with their expertise in how different children learn and what the best activities are that will encourage them to take their learning further" (Young 2014, p. 66). Geography teachers will want to help students develop an understanding of their everyday knowledge within the realms of school and disciplinary subject knowledge.

Knowledge in school is a social construct of the interactions between different groups of people, often subjected to the notion of politicisation of "significant power groups" (Marsden 1989, p. 509). Some of these stakeholders include the state, curriculum planners, academic geographers, geography educators, teachers and students (Chang 2014). Indeed, the school knowledge of a subject discipline is produced by a specialist community where the knowledge is "created by … disciplines … with some rigorous and demanding procedures and practices, put in place to test knowledge claims … to ensure that knowledge created is reliable and truthful" (Lambert 2014).

Young (2011) refers to this disciplinary knowledge as powerful knowledge, as it is "dependable, and testable, taking the learner beyond their experience" (Young 2011, p. 182). Education cannot merely provide access to learning how to read and write; it needs to change the child from a state of daily routine and encounters to one where disciplinary knowledge empowers the child to imagine his or her own future. Maude (2016) proposes five types of powerful geographical knowledge:

1. knowledge that provides students with 'new ways of thinking about the world';
2. knowledge that provides students with powerful ways of analysing, explaining and understanding;

3. knowledge that gives students some power over their own knowledge;
4. knowledge that enables young people to follow and participate in debates on significant local, national and global issues; and
5. knowledge of the world.

Geography disciplinary knowledge inherently includes all five types of knowledge as proposed above. It also responds to the issues raised by geography educators over the last three decades of increased awareness of issues, developments in academic geography and to address knowledge, skills and values that a child needs to engage the future.

In essence, powerful knowledge allows the child to engage new information in a F3 curriculum critically, asks questions about the information based on subject disciplinary knowledge and develops new ways of thinking, powerful ways of analysing, explaining and understanding, takes control of his/her own knowledge and takes part in international debates on issues, thereby enabling him/her to succeed in the world.

In F3 curriculum, powerful knowledge provides a disciplinary lens for inquiry so that the student is able to see the relevance of knowledge to society. Perhaps the question to ask is how can powerful knowledge be taught to students?

As issues and challenges of the world shape the development of academic geography, these developments will eventually find their way into school geography curricula. These changes comprise greater focus on issues of environmental change and globalisation. Besides, developments in education research and the discourse on educational approaches have focused on the value of technology and field-based learning. Gerber (2001) has mentioned the role of fieldwork and technology specifically for a reason. Consequently, this has resulted in higher expectations on teachers to conduct fieldwork or technology-rich lessons.

Real Geography for Students in a Real World

Laws (1989) argues that "real geography depends on good fieldwork". Indeed, fieldwork allows students to develop a holistic and synergistic understanding of geographical issues. While much has been written about fieldwork and its place in geography (Kent et al. 1997), the key value proposition of fieldwork in school geography is in its ability to provide an integration of the theoretical and practical concepts taught in the classrooms through the actual hands-on experience in the field.

Within the F3 curriculum model, geography can be learnt through fieldwork, in allowing students to engage issues based on their geographical understanding of the world. Fieldwork helps to arouse students' interest on an issue, provokes students to identify problems and ask questions, develop perception and appreciation of changing landscapes and enjoy learning about the world they live in (Law 1989). Embedded within these attitudinal objectives of geography fieldwork are the associated knowledge and skills to understand the relationship between the physical features with human landscape and to associate the phenomena which together comprise the geography of an area (Law 1989).

Despite the benefits of fieldwork, the key constraints to using this approach to learning stem from organisation constraint factors such as human resources and time. Bringing a big group of students to the field may be a challenge. There is also the demand on the teachers to do a proper reconnaissance of the site. In addition, teachers have to be mindful of the costs involved in the fieldwork while meeting the budget requirements of their school. As Information and Communication Technologies (ICT) have advanced significantly over the last few decades, there must be ways that ICT can be used to support learning geography of the real world. Favier and van der Schee (2009) suggested that we can have students to investigate real-world problems by combining fieldwork with ICT. We are not suggesting that learning geography with fieldwork or ICT is mutually exclusive, but rather there are opportunities to use both these teaching approaches in a F3 curriculum.

As ICT has advanced rapidly in the past few decades, making searches for geographical facts easily accessible, there is a need to examine how we can get students to move beyond simple acquisition of knowledge to develop their skills to find such facts (Favier and van der Schee 2009). This is especially important when students left unguided can potentially become "uncritical of the information that they find" through Internet search (Parkinson 2013, p. 193). Indeed, we would like to educate children to find information and make sense of what they have found using their geographical ways of thinking—the ideals of F3 curriculum. In other words, students should be able to engage the disciplinary knowledge of geographical thinking to explain, analyse, evaluate, form an opinion and maybe even take action on what they have learnt from the information they have found (Muller and Young 2008). Hence, ICT use can play an important role in helping students learn real geography.

Indeed, ICT can help students to develop knowledge and understanding of "locations and places in order to set national and international events within a geographical framework and to understand basic spatial relationships" (International Geographic Union—Commission on Geographical Education 1992). ICT takes a child beyond the map to an interface where the spatial information can be represented in three dimensions, children can input, manipulate, analyse and retrieve spatial data, so as to identify patterns and relationships between the spatial and non-spatial data. These various Geo-Spatial Technologies enable children to visualise, represent data and understand the real world (Bednarz 2004). ICT can also take the children's learning about the real world beyond their desktop computers to mobile devices such as mobile phones and tablet PCs. There is no denying that children today are already utilising ICT in many aspects of their lives. In a sense, using ICT is an integral part of the real world that they live in. The question then is how do we ensure that the use of ICT for learning geography supports the F3 curriculum. We are not interested in simply providing factual knowledge to our students. Neither are we interested in simply teaching them the skills to search for information. We would like to see students engage the information that they find critically and are supported by their geographical disciplinary knowledge.

What Is This Book About?

While the preceding paragraphs have presented an argument for school geography that must be taught so that it matters to the future of our children, the challenge in any edited book volume is to curate chapters that would come together to provide a compelling narrative, and in this case, an argument for fieldwork and ICT use that will support a F3 curriculum in geography. Lambert and Hopkin (2014) propose that the teachers are key stakeholders within the curriculum-making process. Even if there is a state-mandated curriculum, teachers make decisions about sequencing topics, selecting materials, designing learning activities and assessment, on a daily basis. The argument then is that for geography to be taught well, and subsequently learnt well, the learners' experience, the teachers' choices and the body of knowledge that constitute school geography cannot be independent nor mutually exclusive. In other words, how well the student will learn is dependent on how well a teacher understands the learner's profile, the teachers' subject matter knowledge and the instructional approaches to be used (Lambert and Hopkin 2014).

While the book is organised into four distinct sections for a logical reading flow, the issues surrounding the curriculum-making process in using fieldwork and ICT as ways of teaching geography frame the discussions of the individual chapters. The first section of the book provides the key discussions based on the current understanding about fieldwork and ICT and how the associated learning can be assessed. The second section is dedicated to examples of fieldwork across several country contexts—Malaysia, Thailand and Singapore. The issues of subject matter knowledge, pedagogies as well as the constraints to instruction and assessment are also discussed within these selected chapters. The third section is focused primarily on ICT use for teaching and learning geography, with discussions on the use of Geo-Spatial Technologies, mobile devices and authentic learning. Based on this line-up, we are unable to offer a comprehensive and exhaustive list of country examples or a list of issues for discussion. Instead, the intention of this first edited volume on teaching and learning geography beyond the classroom is to highlight some interesting and keystone examples of issues that will encourage researchers to strengthen the scholarship in these areas. Indeed, this will become a book of examples that could inform further research or innovations in teaching and learning geography beyond the traditional classroom.

Teaching and Learning Geography for the Future

In reflecting about the future of geography, the subject "offers the opportunity to acquire knowledge and skills to see clearer how things are running on planet earth and what we can do differently on a local as well as on a global scale" (Béneker and van der Schee 2015 p. 287). Chang (2015) argues that teaching school geography is not just about teaching a subject, but there is opportunity for the teacher to educate a child and that "[if] we truly embrace the notions of learning about human-environment interaction, space, place, movement and time, then the geography subject allows us to teach a person how to use one's imagination and to be able to think and reason and to decide on how to live based on one's understanding of the environment" (Chang 2015, p. 182).

In Gerber's (2001) survey of geography educators from 32 countries, only 1 out of the 43 respondents for the item reported that fieldwork was not used in teaching and learning of geography. While all the 43 respondents for the item reported that some form of media is used in geography teacher education, the survey did not ask explicitly on the use of ICT. Nevertheless, there is an indication that the place of fieldwork and ICT use cannot be ignored in geography education. The chapters in this book will explore how we can use fieldwork and ICT to teach and learn geography beyond the classroom, with a view to improve geography education for future generations.

References

Bednarz, S. W. (2004). Geographic systems: A tool to support and environmental? *GeoJournal, 60*(2), 191–199.
Béneker, T., & van der Schee, J. (2015). Future geographies and geography education. *International Research in Geographical and Environmental Education, 24*(4), 287–293.
Brundtland, G. H. (1987). *Report of the World Commission on environment and development: "our common future"*. United Nations.
Chang, C. (2014). Is Singapore's school becoming too responsive to the changing needs of society? *International Research in Geographical and Environmental Education, 23*(1), 25–39.
Chang, C. H. (2015). Teaching climate change–a fad or a necessity? *International Research in Geographical and Environmental Education, 24*(3), 181–183.
Favier, T., & Van Der Schee, J. (2009). Learning by combining with GIS. *International Research in Geographical and Environmental Education, 18*(4), 261–274.
Gerber, R. (2001). The state of geographical education in countries around the world. *International Research in Geographical and Environmental Education, 10*(4), 349–362.
International Geographic Union—Comission on Geographical Education. (2016, August). International Charter on Geographical Education. Retrieved August 2016, from International Geographic Union - Comission on Geographical Education: http://www.igu-cge.org/Charters-pdf/2016/IGU_2016_def.pdf
International Geographic Union—Commission on Geographical Education. (1992). International Charter on Geographical Education. Retrieved 2016, from International Geographic Union—Commission on Geographical Education: http://www.igu-cge.org/charter-translations/1.%20English.pdf
Kent, A. (Ed.). (2000). *Reflective practice in teaching*. Sage.
Kent, M. I., Gilbertson, D. D., & Hunt, C. O. (1997). Fieldwork in teaching: A critical review of literature of approaches. *Journal of Geography in Higher Education, 21*(3), 313–332
Lambert, D. (2014). Curriculum thinking, "capabilities" and the place of geographical knowledge in schools', Syakaika Kenkyu (Journal of Educational Research on Social Studies), 81, pp. 1–11.
Lambert, D., & Hopkin, J. (2014). A possibilist analysis of the national curriculum in England. *International Research in Geographical and Environmental Education, 23*(1), 64–78.
Laws, K. (1989). Learning through. In J. Fien, R. Gerbe, & P. Wilson (Eds.), *The Geography Teachers' Guide to the Classroom* (pp. 104–117). Melbourne: Macmllian.

Leggett, J. (1990). *Global warming: the Greenpeace report*. Oxford University Press.
Marsden, W. E. (1989). All in a good cause: History and the politicization of the curriculum in nineteenth and twentieth century England. *Journal of Curriculum Studies, 21*(6), 509–526.
Maude, A. (2016). What might powerful geographical look like?. *Geography, 101*(2), 70.
Muller, J., & Young, M. (2008). Three scenarios for the future - lessons from the sociology of knowledge. Retrieved 20 April, 2015, from Beyond current horizons - technology, children, schools & family: http://www.beyondcurrenthorizons.org.uk
Parkinson, A. (2013). How has technology impacted on the teaching of geography and geography teachers? Debates in geography education (pp. 320). Abingdon: Routledge.
Romey, W., & Elberly, W., Jr. (1989). On being a teacher in the 1990s and beyond. In J. Fien, R. Gerbe, & P. Wilson (Eds.), *The Geography Teachers' Guide to the Classroom* (pp. 407–417). Melbourne: Macmllian.
Young, M. (2011). Discussion to Part 3. In G. Butt (Ed.), *Geography, and the future*. (pp. 181–183). London: Continuum.
Young, M. (2014). Powerful as a curriculum principle. In M. Young, D. Lambert, C. Roberts, & M. Roberts (Eds.), *Knowledge and the future school, curriculum and social justice* (pp. 65–88). London: Bloomsbury.

Chew-Hung Chang is a geography educator serving as the co-chair of the International Geographical Union, Commission on Geographical Education, co-editor of the journal International Research in Geographical and Environmental Education, as well as the President of the Southeast Asian Geography Association. In addition to being a teacher educator, Chew-Hung has published extensively across areas in geography, climate change education, environmental and sustainability education.

Bing Sheng Wu has an expertise in Geographic Information Science and Geo-Spatial Technologies. He teaches across undergraduate to graduate-level courses at the National Taiwan Normal University and has been actively researching on applied aspects of Geo-Spatial Technologies for humanities education.

Tricia Seow As an experienced geography educator, Tricia Seow is involved in the iGeog Steering Committee and Talent Development Programme Fieldwork Trainer, Southeast Asia Geography Association, and the External International Consultant for MA Education Programme. She enjoys fieldwork, and her research interest includes teachers' knowledge and practice, field inquiry in geography and climate change education.

Kim Irvine has worked in the field of hydrology and water resources for more than 30 years. He was awarded the New York Water Environment Association, Environmental Science Award, in 2013. His work has ranged from detailed regulatory studies to capacity building workshops on water quality sampling and assessment for universities, NGOs and government agencies throughout North America and Southeast Asia. His research interests include urban hydrology, water resource management, water quality and applied urban drainage modelling.

Chapter 2
Learning in the Field—A Conceptual Approach to Field-Based Learning in Geography

Diganta Das and Kalyani Chatterjea

Abstract Fieldwork has been considered a hallmark of geographical education by teachers and researchers alike. In the literature review by Kent et al. (1997) on the issue of the effectiveness and importance of fieldwork in geographical education, field studies were found to provide the integration of the theoretical with practical concepts taught in the classrooms. Also, Kent et al. (1997) proposed that fieldwork is commonly accepted as a process that encourages holistic geographical understanding of issues. However, some school teachers commonly conduct fieldwork as field trips where they are in reality just tours or excursions (Chang and Ooi 2008). Students remain largely passive and assume the roles of tourists. Inevitably, these field trips can be less academic, as students are not deeply engaged in the fieldwork process (Brown 1969). On the other hand, properly organized and academically well-articulated field trips can provide students with learning experiences, comparative knowledge, critical understanding as well as skills that are important to an understanding of the world around them (Kent et al. 1997). In practice, many of the fieldwork activities conducted by teachers fall somewhere in the middle on both dimensions. This chapter provides a conceptualization of how an effective field learning experience can be conducted. With a literature review of the range of practices across contexts, the chapter will then uncover steps to identify the issue in the field under study and develop a question, to gather and collect data, to process and reorganize the data, and to reflect and make sense of the information collected. While this simple approach is common to most inquiry-based learning, it provides a clear framework for teachers to conduct meaningful learning of geography in the field.

D. Das (✉) · K. Chatterjea (✉)
National Institute of Education, Nanyang Technological University,
Singapore, Singapore
e-mail: diganta.das@nie.edu.sg

K. Chatterjea
e-mail: kalyani.c@nie.edu.sg

Introduction

Fieldwork allows students to carry out an exploratory task at a field site, outside the classroom to attain some intended learning outcome (see Sim et al. 2005). Through the field, learning takes place with primary experiences of collecting data outside a classroom environment (see Lonergan and Andresen 1988). Cindi Katz (2009, p. 251) observed that fieldwork could be 'a means toward examining the relationships between people and their environments,' and it may help in carefully documenting those relationships and their everyday dynamics (see Das 2014). Fieldwork is hard work, and it contributes to students' overall educational and social skill development. Therefore, Phillips and Johns (2012) observed that fieldwork differentiates the genuine geographer from the not so genuine one. Fieldwork is a fundamental element for learning geography where '[g]eographers learn through the soles of their feet' (Lindsey 1996). Working in the field, geographers collect information and engage themselves in the world beyond the four walls of the classroom (Phillips and Johns 2012). Students and researchers get the opportunity to connect with people and places, develop their field skills, and extend their horizon of geographical knowledge (see Lindsey 1996). Lonergan and Andresen (1988) identified a few primary aims of doing fieldwork in social sciences—first, the field provides opportunities to practice techniques not possible in classroom context; second, field provides opportunities to acquire first-hand knowledge not possible otherwise; third, field studies enhance group learning activities among students; fourth, pursuing fieldwork helps in understanding concepts taught in classroom earlier; and finally, fieldwork enhances higher order learning and arouses concern and appraisal for environment (p. 65).

While geography students learn various concepts and theories in both human and physical geography in a classroom setup, fieldwork helps them to understand the concepts better. Fieldwork, therefore, 'integrate theory with practice.' (see Sim et al., p. 33; Lai 2000). With hands-on experience, fieldwork fuels students' mind with critical thinking and motivates them toward better conceptual engagement. It also helps students getting trained in different ways of conducting fieldwork, understanding the field setting, knowing how to operate equipment (for physical geography fieldwork), and learning various field methods—from observing to interviewing (for human geography fieldwork). Fieldwork has been seen as the bridge between theories and practical concepts (Kent et al. 1997). Former president of Association of American Geographers (AAG) Professor Robert Marston (2005) observed that fieldwork has the power to provide rigorous training and experiences, which 'cannot be replicated in the classroom' (p. 3). Keeping in view the ways fieldwork helps in understanding classroom concepts in geography, the objective of this chapter is to provide a conceptualization of how an effective field learning experience can be conducted through inquiry-based learning and attempts to provide a clear framework to conduct meaningful learning of geography in the field. The next section delves into a literature review of geographical learning through fieldwork, various stakeholders, and their contribution to fieldwork along with a brief discussion on the culture of doing fieldwork in various regions around the world. This section will be followed by examples of doing fieldwork in both human and physical geography courses with discussion on approaches and methods used. These include the ways student deployed classroom concepts to gain insight into the everyday happenings in the field as well as the critical challenges of and negotiations in fieldwork. In conclusion, the chapter summarizes the importance of fieldwork in geographical learning and discusses the effective strategies for conducting successful fieldwork by examining challenges of doing fieldwork faced by both students and teachers at various stages.

Geography and Fieldwork

Geography has long been associated with exploration around the world, voyages, adventures, and journeys (see Couper and Yarwood 2012; Johnston and Sidaway 2004). While the nineteenth century was dominated by the tradition of geography fieldwork as part of a study of nature, everyday landscape and science through exploration, the twentieth century was distinguished by fieldwork in geography as an increasingly academic discipline with scientific observation, regional study, and surveys (Gerber and Goh 2000). In the UK, fieldwork became an integral part of geography courses at universities with Geographical Association (GA) and Royal Geographical Society taking the lead in learning through real-world experiences (Marsden 2000). Further, Marsden (2000) noted that by 1970s, fieldwork was prioritized for environmental studies through the involvement of various stakeholders. Lambert and Reiss (2014) also reiterated the 'unwavering support for fieldwork' by various stakeholders in the UK. In North America, Sauer (1956) argued that fieldwork is fundamental to geography and aggressively campaigned for fieldwork tradition in geography education (see Phillips and Johns 2012). Association of American Geographers (AAG) organizes fieldwork programs for members, especially during its annual meetings. Marston (2005) noted his experience of doing fieldwork as geography students at UCLA and Oregon State University and further elaborated the continuous tradition of fieldwork training in geography education at US universities.

Fieldwork has also been incorporated into geography education especially in Australia, New Zealand as well as many Asian countries. Fieldwork is perceived as an engaging process for creative and independent learning (see Goh and Wong 2000). Fuller (2012) explained the many ways the student learns about the outdoor world through an account of doing fieldwork in various sites of Europe and New Zealand. Similarly, in Southeast Asian contexts, Goh and Wong (2000) argued that learning geography without fieldwork would perhaps be seen as 'deficient' (p. 99). However, the teaching of geography and fieldwork as a teaching practice varies among countries in Asia. India, for example, has a long tradition of geography as a discipline in both school and higher education. However, recent studies (Alam 2014, Tiwari 2012) show that senior high school students lack adequate understanding of basic geography concepts, largely due to under qualified teachers in the discipline. Fieldwork training and related teaching have also been neglected in school geography (see Alam 2015). However, at the University level, fieldwork has been highly incorporated into various courses of the discipline ranging from regional planning and urban geography courses to environmental and physical geography-related courses. Graduate students in Indian universities generally conduct 2 to 3 large field studies ranging from 2 to 3 weeks of 'outside the classroom' experience. In Hong Kong, geography is a 'lively subject' (Kwan 2000: p. 119) with opportunities for experiential learning. The city-state of Singapore, on the other hand, with its tropical surrounding provides students a living lab for doing fieldwork as part of geography learning.

Chang (2012) mentioned the continuous importance of fieldwork in Singapore's geography education and the ways teachers and researchers conduct it. Kho and Parker (2010) noted that in Singapore, doing fieldwork is based on inquiry-based learning process with a well-laid structure to follow. While Rose (1993) noted the increasing trend of gendering in fieldwork—associated primarily with able-bodied men, interestingly, Singapore's geography fieldwork in schools have been organized by a large number of female teachers (see Goh and Wong 2000). While learning through real-world experience in geography has been prioritized in Singapore, teachers do face challenges—ranging from lack of time to plan and prepare for fieldwork to lack of administrative and fieldwork support (Kho and Parker 2010). Chew (2008) noted that at the secondary school level, fieldwork has been accorded lower status in relation to geographical learning and often the fieldwork was conducted through traditional approaches. Nevertheless, fieldwork is cherished and enjoyed by both teacher and student groups for its novelty, and for the element of learning through fun and friendships (see Lai 2000). With new technologies, doing fieldwork in geography has shifted from the traditional approach to new and innovative methods and approaches.

Doing Fieldwork

Fieldwork remains as an integral part of geography education (see Golubchikov 2015; Gold et al. 1991). However, with increasing changes to the theoretical structures and dynamic nature of subdisciplines within geography, methods of doing fieldwork have also transitioned making earlier approaches less usable. The tradition of doing fieldwork has always been associated with physical geography (see Fuller et al. 2000). However, with declining interest in physical geography at school level in the UK, as reported by Fuller et al. (2000), fieldwork is also seen with less enthusiasm (also see Petch and Reid 1988). In physical geography, testing the hypothesis and using new technological tools to measure, weigh, collect, and test materials collected from the field has been increasingly more popular with students and teachers (see Kent et al. 1997). Annual meetings, conferences, and seminars related to physical geography often provide insights into 'the fun or hardships of working in the field and pictures of geographers getting muddy' (Couper and Yarwood 2012, p. 3).

Brown (1969) divided fieldwork into three types. First, 'look and see' fieldtrips which can be less academic and students may not be deeply engaged; second, project-related fieldwork which requires planning in advance with an adequate briefing of the field; and thirdly, fieldwork for collecting primary research data 'with deep interest' (p. 73). Brown (1969) further urged that geographers should develop an integrated approach to teaching geography and provided several approaches for physical geography and do related fieldwork. Fuller et al. (2000) noted that students should be introduced to basics of doing physical geography fieldwork so that they know what to collect and how to collect effectively these when in the field. Pilot fieldwork may perhaps facilitate effective learning for students and larger success of the actual long-duration fieldwork.

Chang and Ooi (2008) observed that doing fieldwork in geography may vary from observation (from a distance and documenting) to active participation. Further, fieldwork may also range from the student being dependent on the instructor to being highly independent in the collection of primary data from the field (also see Kent et al. 1997). Jenkins (1994) noted that with higher student–staff ratios in UK's universities, doing fieldwork with increasing logistics has impacted the effectiveness of fieldwork. Therefore, 13 plausible effective ways of doing fieldwork in geography ranging from not doing fieldwork to focusing on key objectives and making the best use of available resources were proposed in Jenkins' work (Jenkins 1994; p. 145). For effective fieldwork, it is important to have a careful review of the design of the field programme and adequate alignment with the larger course structure and objective (Biggs 2003; in Fuller et al. 2006).

It is tempting to start mapping the taxonomy of fieldwork types proposed by Jenkins (1994) to specific geographical contexts. Fieldwork, specifically in Physical Geography, takes place at a specific location having specific features that is researched on. However, no location is unidimensional and usually offers multiple vantage points that contribute to multi-dimensional knowledge of the environment, human or physical. While some practitioners frown upon Cook's tour type of fieldwork, it is quite usual to pass through information that might only be peripheral with respect to the research subject in focus in a more focused research-oriented field investigation. Such information can contribute immensely toward developing a well-rounded knowledge of the location. This prompts a need for integrating various types of activities while in the field, including observations, field sketching, field data collection of various physical parameters, and the use of relevant IT tools that can enhance all field experiences or provide logistical support. In addition to the specific research objectives, fieldwork should ideally encompass as many observations as possible within the time frame and logistical considerations. This typically may involve peripheral observations and recording of details such as:

(a) What is going on in the area, apart from the specific research topics?
(b) What other environmental factors and processes are salient for the location that has had some influencing the way the specifics are generated?

(c) How are the people living there organizing the physical environment and are themselves organized around the given environment?
(d) How are the future prospects of the place in question shaped by the present, or how do things in the past shape the present aspects?

Observations, therefore, invariably offer a more wholesome knowledge-building method and prepare the researcher to conduct more integrative research analysis, in addition to the range of traditional data collection methods. Human geographers, also, attempt to theorize the everyday human interactions with nature through their observation and ethnographic engagements in the field. In human geography, field instruments increasingly include the use of digital cameras, GoPro, iPads, and even drones as part of visual methods of data collection. It appears then that there is no one single method of data collection that is unique to a context but that an integration of several data collection methods is used.

With well-integrated fieldwork, students can connect to classroom concepts well, leading to higher order learning. Students who were engaged in fieldwork activities felt better connected to the coursework and enjoyed the larger structure of the course (see Fuller et al. 2006). Further, Scott et al. (2006) noted that lecturers also have similar positive attitudes of fieldwork and the effectiveness in better and enjoyable learning for students. For effective fieldwork, Fuller et al. (2006) recommended that it should be integrated into course/modules of study; deeper learning through fieldwork; a less formal environment for conversation between student and teacher during residential fieldwork; and emphasis on hands-on data collection with adequate importance to research design and data analysis. Whether in physical or human geography fieldwork, lecturers need to be involved actively in facilitating the learning for students while in the field—describing the context, explaining in situ happenings, and assisting adequate selection of field techniques (see Fig. 2.1).

Fig. 2.1 A 'hands-on' experience of working in a physical geography fieldwork (Photos courtesy of Professor Kim Irvine)

Learning through fieldwork can be achieved either through structured and guided fieldwork—a traditional approach popular in the 1950s and 1960s. In this traditional approach, teachers take the lead and students participate in a passive mode. Kent et al. (1997) criticized the traditional approach due to its controlled and staff-dependent nature of doing fieldwork. Student-led or self-guided fieldwork, on the other hand, is seen as more autonomous with ownership of the fieldwork learning experience. Here, the student is at the center of learning, acquires new knowledge, and learns the essential elements of doing fieldwork with peers instead of being a passive learner in staff-led field tours (Marvell et al. 2013; Coe and Smith 2010). Self-directing fieldwork can be through either individual investigation or group fieldwork activities. In human geography, participant observation and inquiry-based or problem-based fieldwork are the preferred way of doing student-led fieldwork. These methods provide much-needed autonomy and flexibility to collect data along with deeper involvement by the student or group in field activities to understand in situ geographies and making sense of it with classroom concepts (see Marvell et al. 2013; Das 2014). However, Golubchikov, (2015) cautioned that facilitating student-led independent learning need not be fully 'hands-off'—rather it should be engaged fieldwork that may enhance students' understanding. Further, scholars (such as Farhana 2007; Moser 2008) noted that researchers and students need to be aware of their positionality and power relations while in the field as it may affect the quality of data collected and overall success of the fieldwork. While independent learning through fieldwork is important, the role of teacher/advisor and individual's flexibility, therefore, should not be ignored.

While the majority of the geography fieldwork activities at high school and college levels are done locally, at best around the region with a few days of travel and stay, largely for ease of administrative and logistical preparation as well as plausible nature of the coursework and classroom content, these are geared toward the local and regional understanding of geographical phenomena (see Shurmer-Smith and Shurmer-Smith 2003). In contrast, fieldwork at university level often requires student and teachers to travel far—sometimes taking long-haul flights to understand everyday geographies in other corners of the world. With changing nature of higher education under processes of globalization, internationalization of fieldwork can be seen 'as a kind of global shift' (see McGuinness and Simm 2005, p. 245). Planning of long-haul fieldwork needs plenty of time, administrative and logistical assistance, and meetings between teachers and students to prepare and plan the academic and non-academic requirements. McGuinness and Simm (2005) describe the details of planning the fieldwork module and preparation before commencing the long-haul fieldwork out of Europe. While they recognize that overseas fieldwork could be very tiring for both teachers and students, 'going global' enhances the relationships between research and teaching with relative independent student-led fieldwork (p. 251). Along with the intended learning outcomes, students also experience friendship, bonding over professional and person views, and values not often possible to learn in the classroom.

Fieldwork undoubtedly enhances classroom learning process, makes students aware of geographical phenomena, and assists them to become independent learners. Knowing and learning through fieldwork in geography has been given utmost priority right from the school to the university level in Singapore—with an exciting tropical outdoor lab, the island nation provides exciting opportunities for geographers to do fieldwork (see Das 2014). Keeping up with time and need, Singapore teachers have been successful in continuing the tradition of doing fieldwork as part of geography teaching. At the school level, the fieldwork component is strongly encouraged (Chew 2008). At the university level, fieldwork is an essential module in geography program

However, two significant hurdles that stand in between conducting fieldwork and teaching of geography are (i) shortage of curriculum time and class size and (ii) opportunity to conduct field-based research that generates substantial data to help in-depth research. Teachers are always in a dilemma in solving this problem, leading often to neglect of the fieldwork component. The same problems are often faced in the university. The authors will examine the issues of teacher-led versus independent fieldwork, observation versus participation, local versus overseas fieldwork, issues of curriculum time, and the opportunity to conduct field-based research, through the case examples presented below. These examples are drawn from the teaching experiences of the authors at the National Institute of Education (NIE)/Nanyang Technological University (NTU) Singapore, the institution where both authors work at.

At NIE, students are exposed to fieldwork right from year one to year four, where the learning culminates with a compulsory module on geographical fieldwork. During the first year of their undergraduate course at NIE, students are exposed to various fieldwork methods popularly used by physical and human geographers with some outdoor testing of those methods. During the second year and third year, students are provided with more deep fieldwork knowledge and opportunities for fieldwork in several sites around Singapore. These fieldwork activities—generally organized as day trips—help students to understand the workings of different methods and critical understanding of academic concepts taught in the classroom. During the final year, geography students need to participate in an overseas fieldwork that runs for two weeks.

The authors argue that the examples of the fieldwork can be described by Kolb's (1984) experiential learning model. Kolb's four-stage learning model informs the ways experiences get translated through observation and reflection leading to better understanding and appreciation of (geographic) concepts leading to higher order thinking. Following Kolb's model, Healey and Jenkins (2000) noted that physical geographers may prefer a learning process that is geared toward hypotheses formulation and modeling while human geographers may prefer an adaptive learning process with an inclination to understand various actions and their results (p. 189). Following Kolb's model, geographers can relate contextual learning by doing fieldwork to the theoretical understanding of classroom contents.

Fieldwork in Physical Geography

This section on physical geography showcases three examples of fieldwork carried out by undergraduate students of geography at NIE as part of their course requirements. Two were done locally, and one was conducted overseas. However, all contributed to the final course work. As mentioned in the earlier section, the issues of (i) shortage of curriculum time and class size and (ii) opportunity to conduct field-based research that generates substantial data to help in-depth research have impeded motivation to conduct fieldwork for students. Both teachers in school and university faculty are always in a dilemma in solving this problem, leading often to neglect of the fieldwork component. The examples cited here will show the strategies taken to overcome these hurdles in the physical geography courses at NIE. The first example showcases a field exposure with lecturer presence, which provided immediate on-site supervision for the work done. The second example was done with remote lecturer consultation for a large class, as and when required. This provided more opportunity to the students to take responsibility for the work done and make it more student-centered. In the third example of more senior students, the entire work was student managed, executed, and monitored. The three examples show how field-based research can be incorporated in regular class activities and provide the much-required student-centered learning through inquiry. The scaffolds provided by the faculty either during the field session or prior and post field session aim to expose the students to the Zone of Proximal Development and make them independent learners.

Case Study of Fieldwork at MacRitchie Forest

This fieldwork was done by 50 students of a course on Biogeography in the Central Water Catchment Reserve, one of the reserve forests of Singapore. The entire class was exposed to the concepts and relevant field methods prior to the actual work. The students worked along a small forest stream. The stream section chosen for the work was more than 300 m long, with various important biogeographical as well as fluvial features. The location was difficult to access and required chartering of transport. It was also not possible to visit this place multiple times, due to shortage of free time during regular school hours. The work itself required a long period of time in the field. To overcome these problems, several strategies had to be taken. Working in groups, students were required to contribute data from their own segments and build up a common data set for the entire stream channel. The stretch of the field site was divided into 10 sections, each covered by a group of five students. This division of work ensured collection of a large data set, which was analyzed later in the lab. The entire work had to be completed in three hours, after which the students were ferried back to school for other lessons. So timing of the work was extremely important, and all students were briefed very meticulously on the procedures before the field session. The data collected covered the following aspects:

(a) Environmental data: temperature, R.H., wind condition, heat Index
(b) Stream channel data: stream length, stream width, channel cross section
(c) Streamflow data: streamflow velocity, water temperature, water pH, sediment texture

All these data aimed at providing the students an understanding of the stream and forest environment and the interactions between the forest and the stream. Figure 2.2 shows the field site, students at work, and one of the outputs from their research. The research output ensured that they could integrate all the topics that they learnt in the class about the humid tropical environment, the stream, the forest vegetation, and their interactions.

The field-based research shown can be an example of how the problem of big class size, short teaching hours can be overcome and yet, students can be exposed to field situations, under guidance from the teacher, to ensure adequate support during the exercise.

Fig. 2.2 Groups of students working under guidance, collecting collaborative data

Case Study of Fieldwork at Bukit Timah Nature Reserve

This field research was done by 48 students of the Humid Tropical Environment course. As in the previous case, the size of the class was very large and like any other course at the University, students also had limited time available to do field research. The class was exposed to direct field-based inquiry to investigate how the heavy urban development around the forest affected both the external forest boundaries and the forest interior and to establish zones of impact, based on the extent of changes measured in the environmental parameters such as atmospheric temperature, R.H. wind velocity, soil temperature, radiative temperatures from road and building surfaces. They were then expected to plot the impact zones, using the data collected. The class was, as in the previous case, divided into eight groups, each with six students. But this time, the work involved much longer time and, hence, students could not do it on weekdays, during regular school hours. So the planning involved totally student-centered inquiry, with only scaffolding consultation with the lecturer, on request, from relevant groups. For this, commonly used IT affordances, such as WhatsappTM, were used. The lecturer remained available but remotely, while all groups had to go for the fieldwork on the same day and same time. This was done on chosen weekends but was strictly done during the same time period, as environmental data collected contributed to the common data set, to be used by all groups for their analysis, subsequently. Figure 2.3 shows how the student groups worked and also the maps created after the entire group's data was shared for complete and in-depth analysis.

Again, as in the previous case, pre-fieldwork training in the field methods and the conceptual understanding were of great importance, without which the students would have been stranded. Framework for this field-based inquiry was based on Chatterjea (2010 and 2012), where Stages 1 to 3 were mandatory requirements to progress to Stage 4, when students engaged in fieldwork planning, experimental design, field site monitoring, data collection and subsequent data assimilation, analysis and hypothesis testing. While Stages 1 to 3 were lecturer initiated and scaffolded, Stage 4 was entirely student-driven inquiry. This allowed learner control and learner responsibility. Just-in-time scaffolding through remote connectivity helped in providing them just as much assistance and support as they needed. This integrated classroom and field strategy made students more responsible for the planning and the fieldwork itself, while knowing that the support was present at all times. The extension of fieldwork time beyond school hours helped in allowing students more time at the field, which was impossible to achieve if done during school hours. However, allowing students to work out of school hours without direct supervision required a lot of preparation and training of the students in field etiquettes, safety measures, and general conduct. Training in these is seen as part and parcel of any good field-based learning exercise, and this plan worked toward achieving that.

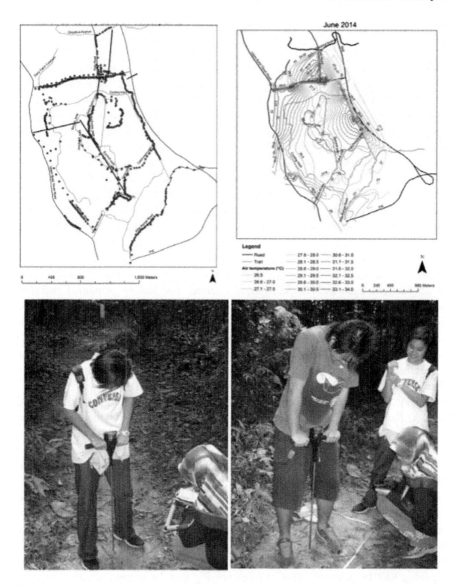

Fig. 2.3 Students working in groups in various parts of a big forest, collecting collaborative data

Case Study of Intense Two-Week Fieldwork Overseas

NIE curriculum requires all graduating students to go for an intensive field-based research outside Singapore in the final year of their study. While the academic justifications are many and are beyond the scope of this chapter, the example of field-based research in physical geography specifically will be discussed in this section. Each year, groups of 15–25 students are taken to various regions outside Singapore, typically for 14–20 days, when they conduct intensive field-based research in varied environments. There have been several modes of operation, with students doing their individual research on their individually chosen topics, which can be quite remote from each other's choice. This involves them going to different places, looking at different things while overseas. This has proven to be a logistical nightmare and allows very limited time for individual research, because the time at overseas locations are limited to a maximum of about 20 days. As a result, the data collected by individual students become very limited and may not allow good analysis.

To avoid this situation, in most years, all students were taken to the same location, where they collected huge sets of different types of data collectively, irrespective of what they finally research on. Subsequently, each student focused on one aspect of the data and analyzed it to produce an individual piece of research. The advantages of this arrangement are: (i) students learn how to work cooperatively in the field, (ii) students are exposed to more types of study elements than their own chosen topic—this enhances their understanding and provides a much broader overview of the world, (iii) the exercise invariably allows building of a very large data set, which can be harnessed for use in any research on the area, (iv) going to the same field location allows the lecturer to be able to accompany them and this opens up scopes for a lot of learning during the journey, again allowing a broad peripheral knowledge development. The above are merely the academic aspects, and the actual learning goes beyond the academic inputs to include developing collective comradeship and collaborative working environment and many more. Even from an academic point of view, such strategies tend to yield impactful learning by offering opportunities for peripheral learning from merely involving in data collection beyond one's own area of research. As mentioned earlier, such involvement in field activities opens the eyes for a broader perspective and finally can add to more wholesome knowledge-building and a more integrative research analysis.

As this strategy involves all students undergoing the same field experience, this kind of fieldwork in physical geography essentially does better with off-the-given-track observations under the guidance of the lecturer. Example of the process could be many. Some examples are given below:

(a) **Stopping along the way to the actual field location**, when some interesting features are observed. This gives the researcher not just an idea that the surroundings are important, but may also provide some clue to the origin and characteristics of the surroundings, even if that may not directly contribute to a chapter in the final write up.
(b) **Maintaining a regular logbook of all field observations** is another way to keep students engaged in this 'additional' observation, as otherwise one tends to forget the details and thus lose out on understanding the big picture. Another additional advantage of maintaining a regular log of activities and observations

is that often discrepancies in field observations can be analyzed and justified by such observations, which might otherwise be missed out.

(c) **The route to the field site and specific locations for observations need to be geo-referenced**. In geography, geo-spatial locations of information are essential, even for extra observations. This provides the much-required spatial significance to all geography research.

(d) **Doing collective fieldwork** requires students to get organized and prepares them to conduct well-coordinated research, within given time frames and resources. While this provides the much-needed training in coordination of actions, it also trains them in more field techniques and at the same time makes them aware of the surroundings. Needless to say that for the work to be effective, a good amount of pre-planning, training is essential and all involved have to be knowledgeable in their individual responsibilities. So the lecturer's involvement in the pre-planning stage is of great importance, as is the requirement of on-site on demand support. For the fieldwork exercises under discussion, use of IT affordances was essential as geo-spatial information and collective data set development were key to effective subsequent field data analysis. Examples of IT usage for this type of fieldwork are discussed in another chapter in this book. The entire exercise was aimed at creating a field environment where students worked in remote locations, collecting massive amounts of data within a given short period. But through collective work process and data sharing facility provided by the mobile application used (NIEmGeo: Chatterjea 2012), students could build up a substantial data repository, which was used by individual students for their own research project. Figure 2.4 below shows some examples of integrated fieldwork by a group of 24 students working at locations that were far from each other, with no visual contact, using NIE mGeo (Chatterjea 2012) and collaboratively creating a database for everyone's use.

Learning through such intensive field-based inquiry helps students to observe features and processes beyond their individual scopes of research, and the knowledge development is always a collaborative effort. While such exercises are usually essential for good research in physical geography, it also contributes to general awareness of the surrounding environment and ultimately enhances learning outcomes. An extract from feedback subsequent to some field-based inquiry is presented below in Fig. 2.5.

The key point that is revealed is that intensive self-directed fieldwork helps make the very important connection between classroom-based concept learning and field-based concept assimilation and a comprehensive knowledge development. This supports the learning framework suggested by Chatterjea (2010, 2012) where knowledge development is only complete after the stages of learning both in the classroom and field are integrated. This integration of knowledge is the ultimate aim of fieldwork in Geography.

Fig. 2.4 Students at work to develop common data sets through collaborative fieldwork

Fieldwork in Human Geography: In Pursuit of Knowing the Unknown

In this section on examples of human geography fieldwork, the authors describe two such cases of fieldwork—beginning with an overseas fieldwork in India for a duration of two weeks and another one-day fieldwork locally at Chong Pang Market in Singapore. The former helped students to use their geographical lenses to observe, participate, and enquire about the everyday geographies of an unknown

Student feedback on Field-based learning session @Sungei Peleoah Kota Tinggi 4ᵗʰ June 2010

	Items	SA	A	D	SD	Total
1	I am clear about how I can plan and design a field based learning session on river processes and features for my students.	4%	96%			
2	I am able to collect and record relevant data, interpret and analyse data to enhance the learning of river processes, features & impact of man's activities on the physical environment.	4%	96%			
3	My knowledge and understanding of river processes, features and impact of man's activities on the physical environment has deepened through the on-site field-based learning session.	13%	87%			

4 My key take-away from the fieldstudy trip is:
  - I learnt how to analyse data and develop deeper understanding of the river features
  - A picture may speak a million words and technology might be able to bring physical landforms closer but nothing can replace seeing and doing it for ourselves. This trip re-affirms the importance of field study in the teaching of geography. Though I might not carry out a similar trip with my students due to constraints, I will definitely do my best to ensure that at least the students have one field trip experience during their upper secondary geographical education.
  - I learnt how to apply geographical skills in the real world
  - The importance of cooperating as a team, and the understanding that we must be prepared to adapt our plans for the environment is dynamic and subject to various changes (e.g. change in water level due to change in climatic conditions). The importance of fieldwork lies not in the result, but the process.
  - Deepened my understanding and knowledge of river processes

5 The best thing for me about this fieldtrip was:
  - The sun, the tan, the activities, the company!
  - To be able to learn from others, fun and getting dirty
  - The first-hand experience of doing what I had only been studying on paper previously. I also have a first-hand experience with the limitations, time control and other matters which will aid in the planning of my future field trip
  - The hands on experience of collecting field data.

Fig. 2.5 Responses from a participant of a field-based inquiry on streamflow process

social and spatial site while the latter helped students to understand a few classroom concepts and relate to the Singapore context. Students were able to understand 'place' through comparative urbanism lens by doing the fieldwork (see Seow and Chang 2016).

Going Overseas—Fieldwork in India

During the fall semester of 2013, two overseas fieldworks were planned as part of the final year undergraduate fieldwork course—a physical geography fieldwork to Thailand and a human geography fieldwork to India. One of the authors led a group to India on the human geography fieldwork. The example will be examined to understand long-durational fieldwork issues, methods, and the application of geographical lenses through 'deep fieldworking' in other geographies.

Following a similar pattern of Kolb's experiential learning, the overseas fieldwork was planned to specifically study urban dynamics and impacts of globalization in the city of Hyderabad in India. Twenty-five students were selected for India fieldwork. While the fieldwork was planned for two weeks during the December of

2013, lectures and planning began six months earlier. The first few lectures were planned to provide an adequate understanding of India's cultural, social, economic background and geographical features of the fieldwork region. Globalization and resultant urbanization being the major theme, theoretical as well as empirical readings were recommended to the students. The next few lectures focused on the application of various human geography-specific methods in relation to Hyderabad. A recce fieldwork was planned in the Little India site of Singapore to understand the workings of the methods, plausible complexities. The recce fieldwork helped in formulating the survey questionnaires and mitigates some difficulties before the actual fieldwork.

On December 6, 2013, the group arrived in Hyderabad after a four-hour flight from Singapore. On the first day, the group began the fieldwork with a visit to a few cultural sites in Hyderabad, and by traveling through the city it provided the students a glimpse of the cultural heritage of the city set against increasing globalizing landscapes. In the next few days, students were able to travel to various sites to conduct their fieldwork by deploying various human geography-specific methods. While the fieldwork was devised to be student-led, it was not hands-off. The instructor maintained a strict vigil on students while working at the sites and provided adequate emphasis on safety issues. At the same time, students were able to observe, conduct surveys and interviews, and write notes. They were in conversations with locals, not necessarily related specifically to their fieldwork. These experiences helped them understand the local context better, workings of geographical concepts, and people's everyday negotiation in the globalizing city. Every night, after dinner, students and the faculty sat down together for an hour to discuss the day's fieldwork, share their experiences, the larger working of the methods and the corresponding difficulties. Based on shared experiences, new strategies were planned for the next day's fieldwork. It is important to mention here that there was immense help from local academics, research assistants, and translators who consistently helped the students in the field, taught them cultural norms of asking questions, assisted them in doing surveys and in the overall efficient collection of data (Fig. 2.6).

Fig. 2.6 In Hyderabad, students were provided with academic learning about the local context with discussions about everyday fieldwork progress and complexities (Photos by the authors)

A Day in Singapore's Chong Pang Market

As part of the coursework of a second-year urban geography module, a one-day fieldwork was organized during the fall semester of 2014. The first few weeks of the module included exposing students to various urban concepts for the better understanding of the urban world. Among them, the concept of urban informality and occupancy urbanism (Benjamin 2008) were discussed through several readings, presentation slides, and student activities during the tutorial sessions. However, it was felt that the contextualization of the classroom concepts will enhance the learning process further. Therefore, a one-day fieldwork was organized with adequate administrative and logistical preparations. Following Kolb's experiential learning model (see Healey and Jenkins 2000), students were provided adequate readings on theories before the beginning of the fieldwork. There were also discussions on plausible methods. Adequate ethical and logistical arrangements were also completed well ahead of the day of the fieldwork.

On the day of the fieldwork, students arrived early in the morning with adequate field materials, notebooks, and cameras. Two methods were selected—observation and visual techniques (Sidaway 2002) based on the nature of fieldwork. While the teacher assisted in preparing the reading materials and the selection of methods, the students were encouraged to be self-guided while doing the actual fieldwork.

There were two reasons for selecting Chong Pang market as the site. Firstly, we wanted to observe and identify elements of informality and occupancy urbanism through everyday practices by residents and the ways it influences liveability. Its location within a mature HDB neighborhood provided us the opportunity to observe the locals lived experiences, their everyday practices, and negotiations. Secondly, Chong Pang market provided us an opportunity to explore a new field site beyond the banality of doing at usual sites such as Little India and Chinatown (Fig. 2.7).

For several hours, students carried out the fieldwork within their small groups through both observation and visual methods. They wrote down their observations in the notebook as part of field diary writing and captured the in situ moments and elements of informality and occupancy urbanism in the real world. By the end of

Fig. 2.7 Everyday informality and occupancy observed in Chong Pang market during the fieldwork (Photos by the authors)

the day, students and the teacher assembled at the nearby community center and shared each other's experiences excitedly and, more importantly, took further notes based on the sharing. The day ended with a lot of fun and banter over lunch. In the following week, students deliberated the ways that fieldwork has helped in a contextualized understanding of the concepts during the regular classroom session. This helped students to learn about doing fieldwork, in a self-directed mode and negotiate field site and complexities of working as a team. While the preparation was onerous, the outcomes made it all worthwhile. In addition to a deeper understanding of the concepts of informality and occupancy urbanism, the students developed stronger relationships, which inevitably contribute to their own social development.

Discussion and Conclusion

The five fieldwork cases presented in this chapter include a broad range of issues of local and overseas fieldwork, teacher-guided and student-centered fieldwork, on-site supervision and remote supervision fieldwork, as well as, physical and human geography fieldwork. These cases have also shown how there are intended versus unintended outcomes.

While local fieldwork takes much less effort to organize than overseas fieldwork, the examples have shown that there is much thought in the design of the fieldwork to ensure that the geographical learning is relevant and authentic. It is not fieldwork for fieldwork's sake but rather guided and grounded in deep geographical understanding. Unlike local fieldwork, overseas fieldwork needs months of planning and preparations before the planned visit (see Irvine et al. 2010). However, overseas fieldwork often opens students' eye to new perspective, inevitably providing them with a refreshed and renewed appreciation of the geographies of a new site. However, planning is often difficult and much thought has to be put into ensure that issues of health, cultural sensitivity, and even safety are well planned for.

While there are basically two main types of fieldwork, teacher-guided and student-led, each has its benefits and limitations. The teacher-guided approach is seen as controlled and staff-dependent (Kent et al. 1997), while the student-led approach is seen as more autonomous, with stronger student ownership of the fieldwork learning experience. It is important to remember that the student-led approach does not mean that teachers can adopt a 'hands-off' attitude but we should be as facilitator to engage students in the learning process, while they observe, conduct surveys, take interviews, and write notes. In either case, there is extensive preparation, whether to develop a questionnaire or to test out the equipment in the laboratory. The ultimate aim of fieldwork is for students to develop deeper geographical understanding. During the course of fieldwork, it is important to have sharing and discussion sessions to help clarify doubts and answer questions to aid students' understanding of geographical concepts and processes. The same goes for on-site versus remote supervision of the fieldwork. While remote supervision provides easier access to all students and on-site supervision allows better control of the data collection process, the teachers' role is to ensure that real geographical learning has taken place.

Fieldwork has been carried out in geography for centuries as it is essential to understanding concepts and theories. While the geography classroom allows students to learn concepts and theories in both human and physical geography, fieldwork provides them with the experiences that help them understand the concepts better. With evolving concepts and increasing development of critical thinking, fieldwork approaches and methods have also been changing to accommodate new dimensions of knowledge. A rather interesting but serendipitous outcome of fieldwork in all the examples shown is that the students have developed stronger friendships through fieldwork. Inevitably they somehow would share in their reflections how they have learned to work with each other better. In light of the rhetoric on twenty-first century skills, fieldwork presents possibilities to truly empower our students.

This chapter began with reviewing the literature on practices of doing fieldwork in geography across several regions of the world and various approaches taken by teachers and instructors, in both physical and human geography. We discussed the effectiveness of these approaches with a critical understanding of reasons behind their adoption and for contextualized learning of geographical concepts and phenomena. Apart from experiences and conceptual understanding, fieldwork also trains students in understanding the field setting, knowing to operate equipment (for physical geography fieldwork) and learning various field methods—from observing to interviewing (for human geography fieldwork). Clearly, fieldwork should be integrated into course/modules of study for deeper learning (Fuller et al. 2006).

Reflecting on popular practices in both physical and human geography fieldwork, we attempted to provide a framework of fieldwork right from the pre-fieldwork arrangement to post-fieldwork analysis and report writing back in the classroom. Through examples of doing fieldwork in Singapore and overseas considerations of doing relatively successful fieldwork is presented.

The chapter has provided some ideas about how fieldwork should be conducted, and the authors would like to encourage geography teachers to conduct meaningful learning of geography in the field. While fieldwork is hard work and comes with various challenges, it also provides opportunities for learning better, to learn to manage time and resources, to learn leadership skills, to negotiate different geographies, to appreciate the culture, and to make long-lasting friendships.

References

Alam, S. (2014). Reorienting undergraduate Geography curricula. *Transactions, Institute of Indian Geographers, 30*(1), 33–43.
Alam, S. (2015). A note on the status of geography teachers in Indian schools. *Geographical Education, 28,* 59–65.

Benjamin, S. (2008). Occupancy urbanism: Radicalizing politics and economy beyond policy and programs. *International Journal of Urban and Regional Research, 32*(3), 719–729.

Biggs, J. (2003). *Teaching for quality learning at University* (2nd ed.). Buckingham, UK: Society for Research in Higher Education & Open University Press.

Brown, B. H. (1969). The teaching of fieldwork and the integration of physical geography. In *Trends in Geography: An introductory survey* (pp. 70–78).

Chang, C. H. (2012). Geography Fieldwork in Singapore. *GeoBuzz,* 11–13.

Chang, C. H., & Ooi, G. L. (2008). Role of fieldwork in humanities and social studies education. In O. S. Tan, D. M. McInerney, G. A. D. Liem, & A. G. Tan (Eds.), *What the West can learn from the East: Asian perspective on the psychology of learning and motivation* (pp. 295–311). Charlotte, NC: Information Age Publishing.

Chatterjea, K. (2010). Using concept maps to integrate hierarchical geographical concepts for holistic understanding. *Research in Geographic Education, 12*(1), 21–40.

Chatterjea, K. (2012). Use of Mobile Devices for Spatially-Cognizant and Collaborative Fieldwork in Geography. *Review of International Geographical Education Online, 2*(3), 303–325.

Chew, E. (2008). Views, values and perceptions in geographical fieldwork in Singapore Schools. *International Research in Geographical and Environmental Education, 17*(4), 307–329.

Coe, N. M., & Smyth, F. M. (2010). Students as tour guides: Innovation in fieldwork assessment. *Journal of Geography in Higher Education, 34*(1), 125–139.

Couper, P., & Yarwood, R. (2012). Confluences of human and physical geography research on the outdoors: An introduction to the special section on 'Exploring the outdoors'. *Area, 44*(1), 2–6.

Das, D. (2014). From classroom to the field and back: Understanding the ways fieldwork empowers geographic learning. *HSSE Online, 3*(2), 14–22.

Kho, E. M., & Parker, W. (2010). Learning beyond the school walls: Fieldwork in Singapore, Grades 1–6. *Social Studies and the Young Learner, 22*(4), 29–31.

Farhana, S. (2007). Reflexivity, Positionality and participatory ethics: Negotiating fieldwork dilemmas in international research. *ACME: An International E-Journal for Critical Geographies, 6*(3), 374–385.

Fuller, I. (2012). Taking students outdoors to learn in high places. *Area, 44*(1), 7–13.

Fuller, I., Edmondson, S., France, D., Higgit, D., & Ratinen, I. (2006). International perspectives on the effectiveness of geography fieldwork for learning. *Journal of Geography in Higher Education, 30*(1), 89–101.

Fuller, I., Rawlinson, S., & Bevan, R. (2000). Evaluation of student learning experiences in physical geography fieldwork: Paddling or pedagogy? *Journal of Geography in Higher Education, 24*(2), 199–215.

Gerber, R., & Goh, K. C. (2000). The Power of Fieldwork. In R. Gerber & G. K. Chuan (Eds.), *Fieldwork in geography: Reflections, perspectives and actions*. Dordrecht, The Netherlands: Kluwer Academic Publishers.

Goh, K. C., & Wong, P. P. (2000). Status of fieldwork in the Geography Curriculum in Southeast Asia. In R. Gerber & K. C. Goh (Eds.), *Fieldwork in geography: Reflections, perspectives and Actions* (pp. 80–99). Dordrecht, The Netherlands: Kluwer Academic Publishers.

Gold, J., Jenkins, A., Lee, R., Monk, J., Shepherd, I., & Unwin, D. (1991). *Teaching geography in higher education: A manual of good practice*. Oxford: Blackwell Publishers.

Golubchikov, O. (2015). Negotiating critical geographies through a "feel-trip": experiential, affective and critical learning in engaged fieldwork. *Journal of Geography in Higher Education, 39*(1), 143–157.

Healey, M., & Jenkins, A. (2000). Kolb's experiential learning theory and its application in geography in higher education. *Journal of Geography, 99*(5), 185–195.

Irvine, K., Vermette, S., & Graber-Neufeld, D. (2010). Developing global scientists and engineers—US undergraduate research experiences on sustainable sanitation and drinking water quality in Thailand and Cambodia. Paper presented at the Southeast Asian Geography Conference, Vietnam.

Jenkins, A. (1994). Thirteen ways of doing fieldwork with large classes/more students. *Journal of Geography in Higher Education, 18*(2), 143–154.

Johnston, R., & Sidaway, J. (2004). *Geography and Geographers: Anglo-American human geography since 1945* (6th ed.). London: Arnold.

Katz, C. (2009). Fieldwork. In D. Gregory, R. Johnston, G. Pratte, M. J. Watts, & S. Whatmore (Eds.), *The dictionary of human geography*. Wiley-Blackwell: Chichester, UK.

Kent, M. I., Gilbertson, D. D., & Hunt, C. O. (1997). Fieldwork in geography teaching: A critical review of literature of approaches. *Journal of Geography in Higher Education, 21*(3), 313–332.

Kolb, D. A. (1984). *Experiential learning: Experience as the source of learning and development*. Englewood Cliffs, New Jersey: Prentice-Hall.

Kwan, T. (2000). Fieldwork in Geography Teaching: The case in Hong Kong. In R. Gerber & K. C. Goh (Eds.), *Fieldwork in geography: Reflections, perspectives and Actions* (pp. 119–132). Dordrecht, The Netherlands: Kluwer Academic Publishers.

Lai, K. C. (2000). Geographical fieldwork as emotionally engaged learning. *Geographical Education, 13,* 25–33.

Lambert, D., & Reiss, M. J. (2014). *The place of fieldwork in geography and science qualifications*. Retrieved from London, UK.

Lindsey, M. (1996). Fieldwork in the undergraduate geography programme: Challenges and changes. *Journal of Geography in Higher Education, 20*(3), 379–385.

Lonergan, N., & Andresen, L. W. (1988). Field-based education: Some theoretical considerations. *Higher Education Research and Development, 7*(1), 63–77.

Marsden, B. (2000). A British historical perspective on geographical fieldwork from the 1820s to the 1970s. In *Fieldwork in geography: Reflections, perspectives and actions* (pp. 15–36). Springer Netherlands.

Marston, R. (2005). The passion for field-based training in geography. *AAG Newsletter, 40,* 3–6.

Marvell, A., Simm, D., Schaaf, R., & Harper, R. (2013). Students as scholars: evaluating student-led learning and teaching during fieldwork. *Journal of Geography in Higher Education, 37*(4), 547–566. https://doi.org/10.1080/03098265.2013.811638.

McGuinness, M., & Simm, D. (2005). Going Global? Long-Haul fieldwork in undergraduate geography. *Journal of Geography in Higher Education, 29*(2), 241–253. https://doi.org/10.1080/03098260500130478.

Moser, S. (2008). Personality: A new positionality? *Area, 40*(3), 383–392.

Petch, J., & Reid, I. (1988). The teaching of geomorphology and the geography/geology debate. *Journal of Geography in Higher Education, 12*(2), 195–204.

Phillips, R., & Johns, J. (2012). *Fieldwork for human geography*. London: Sage Publications.

Rose, G. (2007). *Visual methodologies: An introduction to the interpretation of visual materials*. London: Sage Publishers.

Sauer, C. (1956). The education of a geographer. *Annals of the Association of American Geographers, 46,* 287–299.

Scott, I., Fuller, I. C., & Gaskin, S. (2006). Life without fieldwork: Some staff perceptions of geography and environmental science fieldwork. *Journal of Geography in Higher Education, 30*(1), 161–171.

Seow, T., & Chang, J. (2016). Whose place is this space? Exploring place perceptions and the cultural politics of place through a field-based lesson. *Social Education, 80*(5), 296–303.

Shurmer-Smith, L., & Shurmer-Smith, P. (2003). Field observation: looking at Paris. In P. Shurmer-Smith (Ed.), *Doing cultural geography* (pp. 165–176). London: Sage Publishers.

Sidaway, J. (2002). Photography as geographical fieldwork. *Journal of Geography in Higher Education, 26*(1), 95–103.

Sim, J., Tan, I., & Sim, H. H. (2005). Exploring the use of inquiry-based learning through fieldwork. In C. Lee & C. H. Chang (Eds.), *Primary social studies: Exploring pedagogy and content*. Federal - Marshall Cavendish Education: Singapore.

Tiwari, P. S. (2012). *A note on the teaching of geography in India*. Paper presented at the Contributory paper presented in Symposium on Teaching and Research in Geography in India.

Diganta Das is a human geographer who works on issues of urban development over South Asia. His research interests focus on relations between high-tech space-making and issues of human agency in urban India. He is currently involved in a research project that examines comparative urban liveability of Asian cities.

Kalyani Chatterjea is an avid advocate of physical geography, more specifically geomorphology. Her research focuses on urban geomorphology, rainforest hydrology, environmental change, and channel response. Her special interest includes music and painting and a good holiday destination where one can enjoy wild life and/or mountains.

Chapter 3
Teaching Geography with Technology—A Critical Commentary

Chew-Hung Chang and Bing Sheng Wu

Abstract Information and Communication Technology (ICT) is seen as a way to enhance learning (e.g. Koehler and Mishra CITE 9:60–70, 2009; Rada et al. 1996; Scott 1996), especially in geography (Nguyen et al. 2008; Razikin et al. 2009; Goh et al. 2012). "There is no escaping the web of information technology, and preparing our children to deal with the myriad aspects of this innovation is a task we cannot ignore" (Cheah 1997, p. 140). Some advantages of using ICT for learning include the allowance for self-paced learning, visualization facilitated learning, multi-media learning, constantly updated materials, production of new materials, assessments tailored based on learners' progress and the resource-rich nature of materials from sources such as the Internet. Indeed, learning arises from a constructive process of reflection on the material provided and interacting with it (Farnham-Diggory 1990). There is "growing demand for a 21st century that is independent of time and space, oriented toward goals and outcomes, centered in the student/learner, geared to active, hands-on learning and [the ability] to accommodate differences in skills and language" (Aaggarwal and Bento 2000, p. 4). The authors will provide a conceptual approach to understand how learning geography can be enhanced with technology. Fundamentally, the question that helps us frame our understanding of ICT in geography learning should be "How does ICT help students learn geography better?". Using ICT for learning geography can be understood in terms of how ICT can be used more efficiently, and how to encourage students to learn beyond facts and analyse and apply what they have learnt. With the rise of social media, there is also vast potential for students to curate information and produce knowledge. How then do we develop strategies to ensure that the exponential growth in information is not based on parochial individual naïve theories? Geography as a school subject, the authors argue, engages key affordances of using ICT for teaching and learning. But more importantly, ICT is just a tool and

---

C.-H. Chang (✉)
National Institute of Education, Nanyang Technological University, Singapore, Singapore
e-mail: chewhung.chang@nie.edu.sg

B. S. Wu (✉)
National Taiwan Normal University, Taipei, Taiwan
e-mail: wbs@ntnu.edu.tw

© Springer Nature Singapore Pte Ltd. 2018
C.-H. Chang et al. (eds.), *Learning Geography Beyond the Traditional Classroom*,
https://doi.org/10.1007/978-981-10-8705-9_3

how a teacher chooses the mode of technology, sequence topics, design resources and even assess student learning will be determined by the knowledge and skills required to engage the subject, and in this case, geography.

ICT and Geographical Thinking

In a world where geographical facts, such as the longest river in the world or the most populated city on earth, can be easily searched on the Web and where "students are often uncritical of material sourced online" (Parkinson 2013, p. 193), the challenge is to help students engage the information they have sought through the disciplinary lens of geography. The authors do not intend for the students to accept these facts uncritically, but neither do we want students to become overly suspicious and doubt all sources of information. In an ideal Future 3 curriculum of school education (Young and Muller 2010), students should be able to make sense of the information critically and employ the disciplinary knowledge of geographical thinking to explain, analyse, evaluate, form an opinion and maybe even take action of what they have learnt. Technology should not inhibit learning but should be harnessed to provide the student with the greatest potential to learn geography.

The focus then is on geography learning rather than using ICT. Teaching and learning geography, in turn, is predicated on the definition of what constitutes geography and geographical thinking. Before this discussion digresses too far from the topic, the argument presented here is best demonstrated through an illustration. The vignette created below is purely fictional but encapsulates the ideas about how technology should be used to help students learn geography better.

> Mary, who is an experienced geography teacher has been using technology in her lessons for almost a decade now. Her practice that started out as replication of her teaching materials as PowerPoint slides has evolved into curating multiple sources of online information about a geographical issue, designing an inquiry activity and getting students to use these materials to provide alternative solutions to the issue. Lately she has even begun to use GIS for the lessons because she has a gnawing feeling that using internet sources itself may not be the best way to harness technology for learning geography. When designing her GIS lesson, she first identified the issue to be taught—in this case it was on the erosion of the Changi Point beach, in the eastern part of Singapore. She then studied the aerial photos and maps of the area between 1978 and 2012. She observed the places where there was a distinct retreat in the coastline and places where there were hardly any noticeable erosion. She then converted all the information into layers in the GIS project, each tagged by the year that the information was drawn from. However, she did not stop there. She included a landuse map of the area, embedded photographs taken at the site and she also included a smaller scale map of the area showing the tip of the Malay Peninsula, dominant wind directions in the month of January and July and the major shipping lines off the east coast of Singapore. With these resources included in the GIS project, she then developed the inquiry activity where students were asked to account for the different rates of coastal erosion in the area over the time period and propose ways in which the erosion

may be slowed down. Students then had to test their own hypotheses with the data they have in the GIS project as well as search for additional information from the internet. What is evident here is that Mary wanted her students to employ the geographical concepts such as time, scale, and the human-environment interaction to perform this task, not just to explain, analyse but also evaluate and propose new ways of thinking about the issue that they had not done before.

While the vignette of Mary's use of GIS to help students learn geography is fictional, it illustrates the importance of how technology is used to enable but not restrict the students' learning process. The authors are also not saying that the only way technology can enhance geographical learning is through GIS. Instead, the vignette highlights that technology was used as a tool but that the design of the activity and the geographical thinking behind its design are more important than the technology itself. Indeed, the choice of which technology to use depends on the geographical issue and what the teacher intends for the students to learn.

From Being Consumers to Prosumers of Technology

In the retail market for digital cameras, the products are differentiated into three broad categories based on customer segments—consumers, prosumers and professionals. Consumer-oriented cameras are typically point-and-shoot autofocus devices and offer little in terms of functions but are excellent for convenience and simplicity of use. Professional cameras require sophisticated operations but are able to offer a far more diverse range of functionalities and manual tweaking of settings. Prosumer cameras provide a compromise in that they tend to have expanded functionalities that allow consumers who have more knowledge of the camera functions to capitalize on the advanced features while still allowing for simple functions when convenience is required. The product differentiation in cameras serves as a useful analogy here. Teachers do not need to have highly updated, complex and sophisticated knowledge of technological affordances nor do we want learning experiences with technology to be standardized through wholesale adoption without customization. While technology has expanded the range of information sources and resources for the teacher, what we would like to see will be the clever customization in using the information. In other words, we would like teachers to have some working knowledge of the key technological affordances so that they can make customizations to suit the needs of the curriculum. However, they should still be able to adapt some technological tools for some simple tasks without too much customization.

When examining the role of teachers in the use of ICT for learning geography, the benefits extend beyond instructional uses to that of their own professional development. Indeed, "technology … allows teachers to network and share each other's practices" (Parkinson 2013, p. 193). These could be sharing of informational sources, working in networks or even taking part in activities for the creation of teachers' discourses.

However, one key concern that plagues all teachers in their professional development whether through training or in innovating new practices is that of time, or the lack of. Time available to explore technological affordances and develop meaningful learning activities is constrained both by curricular and extra-curricular demands. This competition for time means that they may not be able to keep up with developments in technological advances, and consequently raises the issue of what the role of a teacher is in using ICT for teaching and learning (Gilbert 2010).

While teachers cannot be expected to be as updated as technology experts in the latest developments within the field, they could do what they do best. They are key curriculum gatekeepers in ensuring that resources students are introduced to or come into contact with will be useful and meaningful in the students' learning. The problem with the exponential growth of technological tools is the amount of information that a student can gain access to. Indeed, some sources of information- and technology-enabled resources the students have access to include:

1. Textbooks and curriculum documents
2. Popular media
3. Print and digital resources
4. Blogs
5. Wikis
6. Videoconferencing
7. Microblogging
8. Chat/RC
9. Digital forum and online communities
10. Social bookmarking
11. Digital photograph sharing
12. Social networking services

These resources have the potential to confuse the students if not used prudently. Of course these resources, which are also available to the teachers, can help them design meaningful learning experience as they organize the materials, sequence learning, share ideas and even collaborate with other teachers to create learning artefacts. Clearly, there are two main ways that technological tools can help the geographical learner—through affordances that support the cognitive endeavour of learning and affordances that support the students' geographical thinking.

Technological innovation has generally changed the way a learner considers content and knowing. Barron and Darling-Hammond (2008) argue that students learn better through authentic learning activities where they use their subject knowledge to solve real-world problems. Indeed, the agency of learning resides with the learners as only then can learning be effective (Hannon et al. 2011). But these affordances can be true of any other school subject. The authors argue that geography and technology have a special relationship in that, the nature of the disciplinary thinking can be enhanced through the use of ICT.

Geographical Thinking with Technology

One of the important objectives of learning geography is to develop knowledge and understanding of "locations and places in order to set national and international events within a geographical framework and to understand basic spatial relationships" (International Geographic Union—Commission on Geographical Education 1992; Wu 2013). A traditional approach to represent space and help students develop geographical thinking is through the use of maps. A map can clearly reflect the two-dimensional spatial phenomenon and patterns and help students retrieve spatial information and construct spatial knowledge intuitively. However, the static representation also has its limits in spatial learning. For instance, it is not straightforward in revealing the dynamics of spatial relation nor in helping people visualize unique physical features on the earth surface. To enhance students' geographical skills and ability to think spatially, various geo-spatial technologies and GIS tools are developed and adopted in geographical education (Bednarz 2004). Nowadays, the emergence of computer technologies and GIS help students by providing opportunities for more hands-on practice so as to familiarize them with spatial concepts and learn to bring real-world geo-objects into digital data. They also learn to use GIS software to represent spatial objects to a 2D or 3D display surface (Meng and Reichenbacher 2005). However, users rely on personal computers or laptops to use existing GIS software, and existing GIS software is designed for professional purposes, and not specifically for geographical learning. In addition to learning the fundamentals of geography, it is essential to explore how individuals become aware of the space around people and the impacts on individuals' spatial understanding (Sui and Bednarz 1999; Bednarz and Bednarz 2004). The use of desktop GIS software is not able to support mobile learning while students explore the space around them. As a result, it is timely to adopt new technologies in geographical education that support the cognitive process of geographical thinking. The following chapters will provide a very brief review of some key geo-technologies, with a view to describe some ways that technology will support the teaching and learning of geography.

Internet and Mobile Devices

The rapid growth of the Internet since the 1990s has dramatically changed the world and accelerated the exchange of information. The development of Web applications and platforms has helped people around the world access and update information in a second. Web technologies allow developers to implement functions originally on desktop applications to Web applications. Users can upload data to the server and share data to others through online forms and Web pages. This trend also changes the development of GIS applications. Online maps such as Google Map or Microsoft Bing Map share spatial data with the public so everyone can easily access maps and get the necessary information to navigate space in the real world. Those maps offer functions such as zoom in/out to dynamically show different spatial information under various scales or provide the best-route searching function to analyse the shortest route between two locations. Digitizing is another prevailing function to create spatial objects such as points, lines or polygons and then to save the information on the server. Web-based GIS platforms such as ArcGIS Server or MapGuide are also developed for deploying Web mapping applications and geo-spatial Web services. Professional GIS software is therefore not the only solution for spatial representation and learning, because users can develop their spatial understanding through the use of Web browsers and online GIS applications. It also implies that users can easily use mobile devices such as laptops or smartphones to retrieve geographical information anywhere, anytime, as long as they are connected to the Internet.

The innovation of mobile devices further expedites the growth of the Internet and forms a new trend of Web 2.0, which allows participation and data sharing from everyone. It changes expert-centred information flow to a new way that collects information from collective intelligence and turning the Web into a kind of a global brain (O'Reilly 2007). The new wave of development has also captured geographers' attention. In geography, individuals act as citizen sensors to share volunteered geographical information (VGI) (Goodchild 2007) to Web-based GIS platforms. A successful example is the OpenStreetMap platform. OpenStreetMap offers a freely editable map to the public so everyone can use mobile devices with global positioning system (GPS) functions to record spatial objectives such as road segment or buildings and then upload locations with related attributes to the platform. Waze, a community-based traffic and navigation platform, shares real-time traffic by collecting traffic data from volunteered drivers. Yelp, a popular platform to share user reviews about local businesses, helps users determine specific type of restaurants around them, for instance. These examples reflect how people receive and use spatial information more often in their daily life. The prevalence of crowd-sourcing content also helps geographers further examine users' spatial understanding and awareness. For instance, Bartoschek and Keßler (2013) developed a Web-based VGI platform and asked students to collect data for specific topics in geography classes. Their research shows that students have benefited through the process of sharing VGI data, and implicitly show how they have expanded their domain knowledge. Wu (2013) used VGI data from students to examine students' understanding of the world and if they used spatial terms properly to describe geographical phenomena. In short, the exposure of the Internet and mobile devices has played vital roles in getting people familiarized with spatial tools and to think spatially.

Virtual Reality and Augmented Reality

Spatial representation and visualization are always essential in geography because the process of constructing spatial cognition relies on how people perceive spatial phenomena through the media or tools that they adopt. The innovation of computer technologies in hardware and software has improved significantly in terms of visualization, and this has stimulated the growth of three-dimensional (3D) technologies. Recently, the technology of virtual reality (VR) has become popular and been broadly adopted in simulation and gaming systems. To reach the best multi-sensory experience, a VR system contains the following key components: 3D display, stereoscopic, headtracked display, hand/body tracking and binaural sound (Earnshaw 2014). Through the help of 3D technologies, a cyberspace can be created and users can discover the virtual world as if they are exploring the real world through computers or VR devices. Second Life is an example that lets users create their own roles and explore the virtual world. People can interact with other online users face to face via their 3D roles. They can also change the virtual environment by adding or removing 3D objects such as buildings or trees, or even changing the terrain on their own. The 3D simulation of the physical world provides a good geographical scenario for users and allows them to experience dynamics of spatial variation (Chen et al. 2013).

Another technology to bridge the cyberspace and real world is augmented reality (AR). AR takes the real world and real environments as its backdrop and inserts computer-generated content (Yuen et al. 2011). After the explosive growth of smartphones, AR is gradually adopted in mobile applications. Another key factor that makes AR popular is the built-in GPS function in smartphones. The global positioning system (GPS) function detects the location of a user's smartphone. Developers can use GPS data to explore surrounding environment and offer certain information, such as restaurants, bus stops or landmarks. This location-based service (LBS) helps cell phone users quickly retrieve spatial information from their surroundings. Since the combination of AR and LBS can easily reveal spatial and temporal information to users, developers and researchers utilize the power of AR and LBS in various fields. Layar is a mobile application that connects digital content with the real world through the use of AR and LBS. Users are able to find icons representing restaurants or banks around them when they observe the surroundings through the camera on their smartphones. Balduini and his colleagues analysed social media data, used the analytical results to offer personalized and localized recommendation of POIs and represented the real-time information by AR and LBS (Balduini et al. 2012).

Interactive Learning by New Technologies

Recent innovation in new technologies has blurred the boundary between the real and cyber world. Users can use fantastic tools or devices to experience virtual space and interact with virtual objects. They can also explore digital information through the lens of reality. The prevalence of mobile devices further expedites the use of new technologies and helps developers create more tools or applications for spatial interaction. Reed and his team created an augmented reality sandbox to show topographical models (Reed et al. 2014). Users are able to reshape the terrain in the sandbox and the corresponding spatial information, including coloured contour lines is re-drawn immediately and projected to represent the new terrain. Users can use hands to mimic precipitation and observe how water flows under various types of terrain. The hands-on practice with new technologies motivates students and helps them develop a clearer understanding of topographic concepts (Woods et al. 2016). Another example is the combination of the Internet, mobile devices and LBS to develop a learning trail (Li et al. 2013). Students can go to designated locations and complete the assigned learning tasks through the observation of the environment and the information offered on their mobile devices. Students will be able to have better training in observation and spatial learning, as a result.

The emergence of computer technologies and GIS has changed the ways of map design and representation. It has also significantly affected how people learn geography and think geographically. New technologies are able to provide a cognitive process that brings geo-objects and their relationships into digital data format and then represents spatial objects on a 2D or 3D display. The use of the Internet and mobile devices also allows users to access spatial information ubiquitously. This blend of real and digital world representations further provides abundant resources for better geographical learning. Since the changes of spatial representation have changed users' experience of spatial understanding, it is essential to adopt new methods for geographical thinking and spatial learning. The question that remains is which technology to use and how can we design meaningful learning with these technologies.

Teachers as Curriculum Makers

When teachers decide how the lessons are sequenced, what activities to design, which resources to choose, what technology to support the learning and how they would assess learning, they are in effect taking part in the curriculum-making process. School syllabus and national curricula do not get translated into the classroom practice miraculously. In fact, the deliberate acts mentioned above can be quite daunting for a beginning teacher. Through experience and continual professional development, a teacher develops deep pedagogical content knowledge (PCK) that will enable him/her to design a meaningful learning activity (Shulman

1986). Indeed, PCK arises as teachers tap into their subject disciplinary training as they interpret the curriculum, find resources that best represent the ideas in the subject matter, and design learning activities that help students bridge the gap in their understanding. The cycle of curriculum, teaching and assessment requires PCK as the learning cannot occur outside this context. Further, an "awareness of common misconceptions and ways of looking at them, the importance of forging connections among different content-based ideas, students' prior knowledge, alternative teaching strategies, and the flexibility that comes from exploring alternative ways of looking at the same idea or problem are all essential for effective teaching" (Koehler and Mishra 2009, p. 64).

In considering the use of ICT in teaching geography, this process of developing the PCK must be informed by some key principles. After examining the key affordances of helping both the learner and the learning of geography, it is evident that the technological pedagogical content knowledge of the teacher is paramount. Technological pedagogical content knowledge (TPACK) is the nexus between what we need to teach, who we are teaching and how we teach. In other words, we are not just interested in using technology but we are interested in matching the learning activity we design to students' needs and the content that needs to be taught.

Central to TPACK (see Fig. 3.1) is the confluence of three types of knowledge—content (CK), pedagogy (PK) and technology (TK). The graphical representation of the intersections between CK, PK and TK is deliberate as these domains of knowledge are not mutually exclusive in a learning activity. Indeed, the way that a teacher can use technology in designing teaching and learning can be described by the intersections of TK with PK and CK, independently and simultaneously. While pedagogical content knowledge (PCK) has been discoursed extensively in describing how teaching and learning are not a purely pedagogic experience independent of the subject matter knowledge to be taught, so is TPACK a reaffirmation that at the heart of using technology for teaching and learning, subject matter knowledge and pedagogical knowledge play important roles. Indeed, TPACK requires consideration for the relationship between the CK, PK and TK domains through issues such as teacher quality, student profile, school culture, resource endowment and of course the topic to be taught. Each situation is different so TPACK provides a framework against which the teacher will ask the key questions of what is to be taught, how do we best teach it and how will technology add value to the teaching and learning of the topic.

The authors argue that TPACK is an operationalization of the interaction between the teacher, the subject matter (school geography in this case) and the student experiences—described as the curriculum-making process by Lambet and Hopkin (2014). In the case of geography, we can consider that the PCK will guide the way the teacher will interpret the curriculum documents and design learning activities for the student. Inevitably, the CK of geography drives the pedagogy and choice of technology use as it requires the teacher to carefully consider what learning activity to choose, what the key concepts are and how they can help students think geographically to take them beyond what they already know.

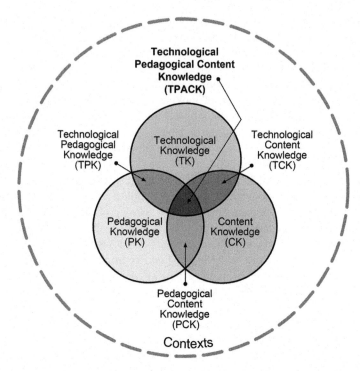

Fig. 3.1 Technological pedagogical content knowledge framework (*Source* http:/tpack.org)

Teachers have to make decisions about "the best of what we know and inducting students into the processes and procedures of how we have come to know it" (Lambert and Hopkin 2014, p. 65) in this curriculum-making process.

Conclusion

As the vignette of Mary has demonstrated how her skilful design of her lesson is informed by her own geographical knowledge, an understanding of the affordances that technology provides and the recognition of the students' learning profile, the overarching principle in using technology for geographical learning emerges. In a simple process of elimination, we can start asking ourselves which of the three principles if removed will make it impossible for students to learn geography with technology. Unfortunately, the answer will be technology. This pessimistic outlook is perhaps flawed in that we have always considered that ICT must be something new and different from what we have been doing. However, technological advancements arise because there is a need for certain functionality that we do not yet have.

An interesting observation that the authors would like to make on how there is no escaping from technology comes from science fiction. In the science fiction franchise of the Star Trek™ series, the producers often have to dream up some fantastic and yet functional technology for their characters in the far future. Many of these imagined technologies have become reality today (and a whole lot more have not). In deciding what future technology to "invent," the functionality is determined by the requirements of the plot. One of the authors' favourite invention is the PADD™—Personal Access Display Devices, first introduced in the late 1980s in the series Star Trek—The Next Generation™. The series required a futuristic device that could replace the clipboard, and with constraint on costs, they invented a touch screen, flat display that could fit into the characters' hand (Forseman 2016). While one may argue that today's portable tablet computers might have been inspired by these props, the point that the authors are making here is that if there is a compelling functionality that needs to be invented, it will be invented. Technological innovation arises because someone needs something to be invented.

By reflecting on the history of technological advancement, we realize that there have been technological innovations that would just provide incremental improvement to way things are done and those that are so disruptive in nature that they replace the way that things are normally done. One example is the Guttenberg Press that replaces the need for painstaking hours of calligraphic diligence in producing texts. Another is the smartphone. Cellular phones no longer function purely as a telephone but act as a communicator, a browser, an aggregator of information, a media producer and even a data collection device. While we can argue that each of these functions has dedicated devices that could rival a cellular phone, the undeniable fact is that our children are very savvy with technology cannot and will not be able to get through a day in their lives without the smartphone. This realization negates the argument that one does not need technology to learn geography. In fact, ICT is an integrative part of our students' reality. Indeed, there is often the concern that in the absence of proper education programmes that target cyber wellness, our students may lose the perspective that geographical issues occur in the real world. It is with this perspective that this book deliberately incorporates the importance of fieldwork for learning geography. Consequently, the examples of technology use, integrated with fieldwork for geographical learning, have been included in this book.

References

Aggarwal, A., & Bento, R. (2000). Web-based education. In A. Aggarwal (Ed.), *Web-based learning and teaching technologies: Opportunities and challenges* (pp. 2–16). Hershey PA: Idea Group.

Balduini, M., Celino, I., Dell'Aglio, D., Della Valle, E., Huang, Y., Lee, T., et al. (2012). BOTTARI: An augmented reality application to deliver personalized and location-based recommendations by continuous analysis of social media streams. *Web Semantics: Science, Services and Agents on the World Wide Web, 16,* 33–41.

Barron, B., & Darling-Hammond, L. (2008). *Teaching for meaningful learning: A review of research on inquiry-based and cooperative learning.* Book Excerpt. George Lucas Educational Foundation.

Bartoschek, T., & Keßler, C. (2013). VGI in education: From K-12 to graduate studies. In Crowdsourcing Geographic Knowledge (pp. 341–360). Springer, Dordrecht.

Bednarz, R. S., & Bednarz, S. W. (2004). Geography : The glass is half full and it's getting fuller. *The Professional Geographer, 56*(1), 22–27.

Bednarz, S. W. (2004). Geographic systems: A tool to support and environmental ? *GeoJournal, 60*(2), 191–199.

Cheah, Y. M. (1997). Shaping the classrooms of tomorrow: Lessons from the past. In J. Tan, S. Gopinathan, & W.K. Ho (Eds.), *Education in Singapore, A book of readings* (pp. 129–144). Singapore: Prentice Hall.

Chen, M., Lin, H., Hu, M., He, L., & Zhang, C. (2013). Real-geographic-scenario-based virtual social environments: Integrating geography with social research. *Environment and Planning B: Planning and Design, 40*(6), 1103–1121. https://doi.org/10.1068/b38160.

Earnshaw, R. A. (2014). *Virtual reality systems*. Elsevier Science.

Farnham-Diggory, S. (1990). *Schooling, the development child*. Cambridge: Harvard University Press.

Foresman, C. (2016, September 11). *How Star Trek artists imagined the iPad… nearly 30 years ago.* Retrieved September 30, 2016, from Ars Technica: http://arstechnica.com/apple/2016/09/how-star-trek-artists-imagined-the-ipad-23-years-ago/.

Gilbert, I. (2010). *Why do I need a teacher when I've got google*. London: Routledge.

Goh, D., Razikin, K., Lee, C., Lim, E., Chatterjea, K., & Chang, C. (2012). Evaluating the use of a mobile annotation system for geography education. *The Electronic Library, 30*(5), 589–607.

Goodchild, M. (2007). Citizens as sensors: The world of volunteered geography. *GeoJournal, 69*(4), 211–221.

Hannon, V., Patton, A., & Temperley, J. (2011). Developing an innovation ecosystem for education. Cisco White Paper December.

International Geographic Union—Commission on Geographical Education. (1992). International Charter on Geographical Education. Retrieved 2016, from International Geographic Union—Commission on Geographical Education: http://www.igu-cge.org/charter-translations/1.%20English.pdf

Koehler, M., & Mishra, P. (2009). What is technological pedagogical content knowledge? *CITE, 9* (1), 60–70.

Lambert, D., & Hopkin, J. (2014). A possibilist analysis of the national curriculum in England. *International Research in Geographical and Environmental Education, 23*(1), 64–78.

Li, Y., Guo, A., Lee, J. A., & Negara, G. P. K. (2013). A platform on the cloud for self-creation of interactive trails. *International Journal of Mobile Learning and Organisation, 7*(1), 66–80.

Meng, L., & Reichenbacher, T. (2005). Map-based services. In L. Meng, A. Zipf, & T. Reichenbacher (Eds.), *Map-based mobile services: Theories, methods and implementations* (pp. 1–10): Springer.

Young, M., & Muller, J. (2010). Three educational scenarios for the future: Lessons from the sociology of knowledge. *European Journal of Education, 45*(1), 11–27.

Nguyen, Q. K. (2008). TagNSearch: Searching and navigating geo-referenced collections of photographs. *International Conference on Theory and Practice of Digital Libraries* (pp. 62–73). Berlin, Heidelberg: Springer.

O'Reilly, T. (2007). What is Web 2.0: Design patterns and business models for the next generation of software. *Communications & Strategies*, (1), 17.

Parkinson, A. (2013). How has technology impacted on the teaching of geography and geography teachers. *Debates in Geographical Education*, 193–205.

Rada, R., Rimpau, J., Bowman, C., Gordon, J., Henderson, T., & Sansom, T. (1996). World wide web activity and the university. *Educational Technology Research and Development, 36*(5), 49–51.

Razikin, K., Goh, D. H. L., Theng, Y. L., Nguyen, Q. M., Kim, T. N. Q., et al. (2009, October). Sharing multimedia annotations to support inquiry-based learning using MobiTOP. In *International Conference on Active Media Technology* (pp. 171–182). Berlin Heidelberg: Springer.

Reed, S., Kreylos, O., Hsi, S., Kellogg, L., Schladow, G., Yikilmaz, M., et al. (2014). Shaping watersheds exhibit: An interactive, augmented reality sandbox for advancing earth science. Paper presented at the *AGU Fall Meeting Abstracts*.

Scott B. D. (1996) Teaching the net: Innovative techniques in Internet training. Paper presented at the *Annual Computers in Libraries Conference* (Washington DC, February 1996).

Shulman, L. (1986). Those who understand: Knowledge growth in teaching. *Educational Researcher, 15*(2), 4–14.

Sui, D. Z., & Bednarz, R. S. (1999). The message is the medium: Geographic in the age of the Internet. *Journal of Geography, 98*(3), 93–99.

Woods, T. L., Reed, S., Hsi, S., Woods, J. A., & Woods, M. R. (2016). Pilot study using the augmented reality sandbox to teach topographic maps and surficial processes in introductory geology labs. *Journal of Geoscience Education, 64*(3), 199–214.

Wu, B. S. (2013). Developing an evaluation framework of spatial understanding through GIS analysis of Volunteered Geographic Information (VGI). *Review of International Geographical Education Online, 3*(2).

Yuen, S., Yaoyuneyong, G., & Johnson, E. (2011). Augmented reality: An overview and five directions for AR in education. *Journal of Educational Technology Development and Exchange, 4*(1), 119–140.

Chew-Hung Chang is a geography educator serving as the Co-chair of the International Geographical Union, Commission on Geographical Education, Co-editor of the journal International Research in Geographical and Environmental Education, as well as the President of the Southeast Asian Geography Association. In addition to being a teacher educator, he has published extensively across areas in geography, climate change education, environmental and sustainability education.

Bing Sheng Wu has an expertise in Geographic Information Science and Geo-Spatial Technologies. He teaches across undergraduate to graduate level courses at the National Taiwan Normal University and has been actively researching on applied aspects of Geo-Spatial Technologies for humanities education.

Chapter 4
The Importance of Assessing How Geography Is Learnt Beyond the Classroom

Chew-Hung Chang and Tricia Seow

Abstract Finding out if someone has learnt what you intend for them to learn remains a key issue in geography education. It is certainly not as simple as looking at a student's performance through pen and paper examinations. To extend this concern to the future, we ask if geography educators will be contented that a child has only learnt geographical knowledge—arguably the most easily assessed domain through traditional assessment formats. How do we evaluate if our student is developing skills that will help him/her engage the environment that he/she is living in better? If we aim for attitude and behavioural changes, how will we know that our students have become better custodians of our common environment? Good assessment practices in geography will allow the teacher to determine how well they are teaching and how well the students are learning. Coming back to the theme of the book, the challenge is to extend good assessment practices beyond the conventional classroom to the use of ICT as well as in the field. The twofold challenge of determining what is good assessment and how to extend it beyond the classroom will be discussed in this chapter. By providing the key dimensions to evaluate what good geographical assessment entails, and then extending these to assessment in the field and in using ICT, the reader will be able to engage the issues raised in the subsequent chapters.

Assessment in the Borderless Geography Classroom

Teachers are interested to conduct good assessment because they want to know how well they are teaching and how well the students have learned (Voltz et al. 2010, p. 116). If a student does well in the assessment, it is more likely that the teaching

---

C.-H. Chang (✉) · T. Seow (✉)
National Institute of Education, Nanyang Technological University,
Singapore, Singapore
e-mail: chewhung.chang@nie.edu.sg

T. Seow
e-mail: tricia.seow@nie.edu.sg

© Springer Nature Singapore Pte Ltd. 2018
C.-H. Chang et al. (eds.), *Learning Geography Beyond the Traditional Classroom*,
https://doi.org/10.1007/978-981-10-8705-9_4

process has been successful at helping the student learn better. Of course, doing well in class tests or assignments need not be an indication that good teaching and learning has taken place. Indeed, it could also be that the material is too easy for the students or that they are remarkably capable and have done well despite how poorly they have been taught.

Putting the efficacy of teaching aside, assessment is useful to find out what students know. This allows the teacher to determine which student needs more help or which topics need to be revised. This information from assessment therefore lets teachers develop a profile of each student's learning, allowing him/her to employ differentiated instruction or even remediation. In addition, assessment allows teachers to determine the best timing in the curriculum to use different types of resources or design different learning activities.

Indeed, assessment helps us determine what students have learned and what we expect students to learn from what we have taught. However, we must also be mindful that assessment serves a practical function of reporting students' progress to stakeholders such as school administration and parents. This requires assessment to be reliable and valid, such that the performance of individuals should be consistent regardless of when the assessment is used and it should measure the intended learning outcomes.

If assessment must have "consequential validity" (Lambert and Balderstone 2000, p. 334), a concept first developed by Professor Caroline Gipps (Gipps 1994), then the way teachers gather information about students' learning must be designed so that it has implications for improving learning. In geography, the assessment of learning outcomes extends beyond the traditional classroom. Geography is a discipline that has traditionally been linked to the exploratory tradition (Sauer 1956; Stoddart 1986; Driver 2001), and "fieldwork has always been central to the enterprise and imaginary of geography" (Bracken and Mawdsley 2004, p. 280). In addition, the visual traditions of geography are well documented (Cosgrove 2008; Rose 2016), which in recent times have been extended to ICT and GIS technologies. As teachers explore the use of ICT and field-based activities for geography learning, there must be a consideration beyond instructional design to assessment that has consequential validity. This is vital, so that we can draw sound conclusions about what the student has learnt. Good assessment tasks should be closely related to the intended learning objective, but what is a valid test for one learning objective may not be appropriate for another. Indeed, good assessment tasks should be aligned to the learning outcomes for the curriculum.

Assessment as an Integral Part of the Geography Curriculum

Assessment should not be something left to the end of a lesson or a learning activity. It should be planned well before the students start a learning activity (Lambert and Balderstone 2000, p. 339). While planning for assessment can occur at a state or district level, the school level and the classroom level, with varied intentions, the end goal does not deviate from the desire to see our students improve in their learning. The need to plan assessment deliberately even before the learning occurs puts it squarely within the curriculum-making process.

While education researchers have discoursed on the notion of curriculum as a linear or cyclical progression of planning, delivery and evaluation of an intended course of study (e.g. Tyler 1949), it has also taken on multi-perspective orientations (Miller 2005) which can be considered "an extraordinarily complicated conversation"(Pinar et al. 1995). A curriculum is not just a planned sequence of work that guides instruction, but it is "a complex social and political construction built on understandings of different groups in society (including teachers) based on the past and present and with an imagination for the future" (Goodson 1997). Indeed, a teacher's role in the curriculum-making process is not just to read off a checklist of activities suggested by a curriculum document but to make sense of the curricular requirements while actively making decisions about what to teach, how to teach and how to assess.

Lambert and Hopkin (2014) propose that the curriculum is a product of a dynamic interaction between three domains of the child, the way the subject is taught, and the broader social purposes of education. The interaction between the three domains is shown in Fig. 4.1.

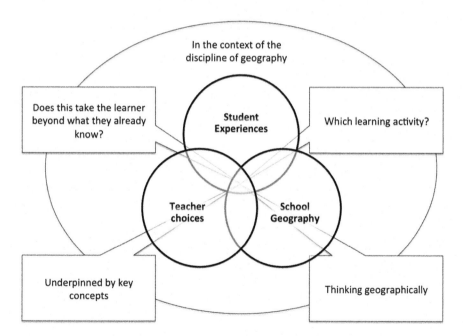

Fig. 4.1 Curriculum making in geography (Lambert and Hopkin 2014)

In the case of geography, we need to consider the student's "perceived needs and our understanding of learning". To this end, teachers then curate "the best of what we know and inducting students into the processes and procedures of how we have come to know it" (Lambert and Hopkin 2014, p. 65). The teacher plays an integral role in the curriculum-making process. Indeed, the Boolean intersection between the three domains of curriculum making in geography requires the teacher to carefully consider what learning activity to choose, what the key concepts are and how they can help students think geographically to take them beyond what they already know. While it is tempting to develop a checklist to guide teachers in this curriculum-making process, this reductionist approach defies the intention to empower learners to engage in the knowledge acquisition process critically.

In balancing the lofty theoretical ideals of curriculum with the pragmatics of enabling the teacher to understand this process and to implement it, there is a need to provide some semblance of a guideline without constraining the teacher with too many instructions. To this end, the curriculum-making process can be guided by task design, first proposed by Doyle and Carter (1984). Tasks are defined as classroom events that give form to curriculum, where teachers create structured activities that require students to engage the subject matter (Doyle and Carter 1984). The task designed is a confluence of an interpretation of the curriculum, an understanding of the subject matter, selection of instructional approaches and delimiting assessment procedures, which constitute the learning activity. Some examples of tasks given by Doyle and Carter (1984) include performing an experiment, giving responses to watching a video and even writing an essay. This has very similar features to performance assessment tasks, which are integral to the Understanding by Design (UbD) approach (Wiggins and McTighe 2005).

The UbD approach to designing a unit of study focuses on unpacking an "enduring understanding" statement that, in our opinion encapsulates the key concepts of a unit of learning, informed by the disciplinary thinking of the subject. For instance, an enduring understanding statement on the topic of resource management like "how humans use resources has an impact on the environment and the future" must necessarily draw on the geographical concept of the human–environment interaction. In other words, the teacher makes decisions about the design of the lesson and the corresponding assessment based on this understanding.

Also known as the backward design approach, UbD is focused on developing the assessment tasks before the instructional activities. Indeed, once the key understanding of a unit of learning has been identified, a performance assessment task is designed first. While the authors are not suggesting that Doyle and Carter's task design is the same as the UbD design approach, there is convergence between the two approaches on designing assessment that has consequential validity. Within the context of school geography, we are arguing that assessment both within and beyond the traditional classroom need to help students learn geography better.

Assessment Framed Around Learning Outcomes—Disciplinary Knowledge in Geography as the Key

A well-designed assessment task in the field and in using ICT should clarify the learner's thinking process and should avoid testing for irrelevant content (Wiggins and McTighe 2005). As discussed above, it should also be purposeful. Teachers designing assessment need to be clear about what we want our learners to be able to demonstrate and use this as a basis to determine what and how to design a learning activity to support this. The alignment of the assessment intentions and the activity design will allow learners to have clarity in the criteria used to assess them and also allow the teachers to determine beyond performance to uncover why they have performed this way (McTighe and Self 2002).

Drawing on Bloom's original taxonomy of learning objectives, Krathwohl (2002) has provided a taxonomy that looks at both the knowledge domains and cognitive processes (see Table 4.1).

This categorization of knowledge domains and cognitive processes to be learnt will help us determine if students **develop knowledge, skills, values and action to engage and learn geography**. However, we need to consider the curriculum content for geography so that we can design assessment that will enable students to demonstrate the performance of those outcomes by specifying the criteria to be evaluated (Cohen 1995).

Geography is concerned with asking questions of:

1. Where is it?
2. What is it like?
3. Why is it there?
4. How did it happen?
5. What impacts does it have?
6. How should it be managed for the mutual benefit of humanity and the natural environment?

(International Geographic Union—Commission on Geographical Education 1992).

Table 4.1 Krathwohl's revision of Bloom's taxonomy

	Cognitive processes					
The knowledge dimensions	Remember	Understand	Apply	Analyse	Evaluate	Create
Factual						
Conceptual						
Procedural						
Metacognitive						

Source Krathwohl (2002), p. 216

In order for the learning outcomes to guide the assessment design for consequential validity, these learning outcomes must be clearly defined within the subject discipline of geography. The questions about geography listed above are drawn from the first International Charter on Geographical Education. A closer examination of that document also shows several other ways of looking at how to determine what we want our students to learn in geography.

In terms of geographical concepts, the following were listed:

1. location and distribution
2. place
3. spatial interaction
4. region
5. people–environment relationships

In terms of skills, the following were included:

1. identifying questions and issues
2. collecting and structuring information
3. processing, interpreting and evaluating data
4. developing generalizations
5. making judgements
6. making decisions
7. solving problems
8. working cooperatively

The skills outlined above were also listed in relation to the type of information and data that geography students have to work with:

1. maps
2. diagrams
3. tables
4. graphs
5. pictures
6. symbolic data
7. quantitative data
8. verbal information

(International Geographic Union—Commission on Geographical Education 1992)

While the authors are not proposing that all learning outcomes should be designed according to one of these schemas of describing geographical learning, these ways of thinking of learning geography should inform how we design assessment both within and beyond the classroom. Indeed, geography should be concerned with "the study of Earth and its natural and human environments. Geography enables the study of human activities and their interrelationships and interactions with environments from local to global scales" (International Geographic Union—Commission on Geographical Education 2016, p. 4). This affirmation in the new International Charter on Geographical Education provides the background to the argument for consequential validity in this chapter.

Reflecting on High-Stakes Paper and Pen Examinations

High-stakes testing drives instruction in the classroom (Voltz et al. 2010, p. 114). Unfortunately, when the word assessment is mentioned, there is the inevitable image of pen and paper tests. If we use summative assessment to ascertain how much individuals have learnt, then formative assessment is not a mutually exclusive or independent endeavour from pen and paper tests. Assessment can be integrated within instruction for learning (Hagstrom 2006). While some have considered classroom assessment to be weak (Black 2000), there is vast potential in developing assessment for learning. While there is prevalent practice of formative assessments in the form of class quizzes, reflection papers, posters or project work within the classroom (Chang 2014), these practices need not result in better skills nor knowledge about the topic to be learnt. Indeed, the students may not even see much relevance of these classroom assessment tasks to their daily lives apart from scoring well in the examinations (Chang 2014). Nonetheless, there is still a place for summative assessment as "it should inform the process before, during and after teaching has occurred" (Voltz et al. 2010, p. 116).

In considering how assessment could support learning geography beyond the conventional classroom, there has been an interesting development in Singapore where level-grading and field-based geographical investigation has become a key component of the high school geography subject over the last decade. Since 2007, high school geography in Singapore has adopted a "levels marking" approach for part of the national written examination paper (Singapore School Examinations and Assessment Board 2010). The objective was to measure students' ability to discuss and evaluate geographical problems beyond simply describing or explaining. Before this, the national examination for geography required students to answer 40 multiple choice questions and four structured essay questions in which marks were awarded for relevant and accurate points raised for the structured essay questions. The 2007 review which deviates from point-allocation grading supports students in engaging " the challenges of an increasingly globalised world … [and] to promote critical and creative thinking skills, and to nurture problem-solving and independent learning abilities in students" (Sellan et al. 2006).

A further development in 2014 was the introduction of a geographical investigation component (Singapore Examinations and Assessment Board 2014) in which students have to answer questions about how field-based geographic investigation can be conducted, how information and data can be collected, organized and analysed, presented and what they can conclude based on their findings. The field-investigation was designed based on Roberts' (2003) cycle of inquiry. The assessment innovations described here are in alignment with the vision of the

International Charter on Geographical Education in which school geography is seen as a way to prepare our children to engage in the global issues of our time.

Although this is an example in Singapore and may not be similar to other countries' and states' geography education contexts, the chapters in this book are selected to discuss examples across the region of how learning geography can take place on a barge or by using mobile technology to tag locational and geographical data, for instance. It is through examining each individual context that we should consider assessment.

Geographical Knowledge and Morality

The authors also suggest that the forms of assessment that prevail in high-stakes summative assessments are inadequate because they privilege the cognitive and skills domains and ignore behavioural and attitudinal outcomes. The following quotations come from two students in Singapore in response a question about whether what he/she has learnt in the geography classroom has shaped his/her environmental behaviour:

> I don't think most people would bring home what they actually discussed about. And some people would forget about it. Some people actually take down notes to study for the exams. I think after the exams, everybody would just (pause) yeah, forget about it.
>
> Actually, I think exams are very effective of making us remember things. But erm…but (if) you remember, do you do it? I don't, you know. I remember, I know everything. I know things that I'm supposed to remember, but I don't…(long pause).
>
> (Chang 2014)

Considering behavioural and attitudinal dimensions in assessment draws us into ongoing debates related to the purpose of a geographical education. Lambert and Balderstone (2000) argue that the teachers' role is a moral one and teachers cannot do their work without considerations of ethics. However, this stance is not universally agreed upon. Within environmental education for instance, there are researchers who believe that teachers should seek to develop independent thinking and critical thought (Jickling 1992; Aldrich-Moodie and Kwong 1997) in their classrooms and that advocating for values education and social change for the environment (as suggested by Fien 1993; Huckle 1985; Morgan 2012) is anti-educational (Williams 2008).

Geography educators appear just as divided over the ethical dimensions of their work. For instance, studies have noted that teachers are reluctant to advocate for the environment in their lessons (Tomlins and Froud 1994; Cross 1998), and are more comfortable adopting a neutral stance while presenting a range of different viewpoints about environmental topics (Cotton 2006). In contrast, other researchers have found that teachers feel responsible for promoting environmental attitudes and behaviours (Ballantyne 1999). In fact, Ho and Seow (2015) have found that the same geography curriculum in Singapore is interpreted and enacted differently by teachers based on their beliefs about the purpose of geography education.

Geography educators have to grapple with such questions about disciplinary knowledge in geography and its moral implications when designing learning activities and assessments that have consequential validity. The form of assessment that they select will differ markedly based on where they stand on the issues. For instance, Seow (2015) suggests that teachers planning for fieldwork need to first consider carefully the purpose of the field-based experiences they provide for students since this affects the form of assessment that teachers should design. Teachers who wish to emphasize geographical knowledge may choose to assess students' ability to apply geographical theories learned in the classroom to make sense of the dynamic processes they observe in the field and to reconcile these with anomalies observed in the real world. Those who prefer to focus on developing geographical skills may develop assessment around students' competencies in procedural field-based skills. In contrast, teachers who are guided by behavioural or attitudinal dimensions may seek to evaluate the types of actions that students choose to take to resolve particular problems in the field and pay close attention to the reasons undergirding these actions.

Conclusion

"Assessment is not an exact science and we must stop presenting it as such" (Gipps 1994, p. 167). While the chapter has provided multiple perspectives on why assessment is important, how it can be designed beyond the traditional geography classroom and how it should take the student beyond just cognitive engagement to developing a whole child, it is not difficult to see how complex the issue of assessment is. Rather than prescribing how we should all assess geography learning, both inside and beyond the classroom, the chapter was written with the intentions to provide different ways of approaching the issue.

At the core of our discussion, we are interested to find out if a student has learnt what we have intended for them to learn. The concept of consequential validity helps us design assessment that will help improve student learning in geography. While pen and paper examinations remain a mainstay of the education process, alternative but not mutually exclusive ways of assessing students' geographical learning have been examined. Curricular alignment is one way that we can develop formative assessment both within and beyond the classroom, but we also want to see our students developing skills that will help them engage the environment that they are living in. We have provided some key discussion points that we are sure will frame the discourses in the subsequent chapters in this book. "Assessment is today's means of understanding how to modify tomorrow's instruction" (Tomlinson 2014, p. 17). Nevertheless, assessment should be considered as an integral part of the curriculum-making process, for while it helps us determine what students have learnt or how effective our teaching has been, it also helps us design instruction (especially beyond the traditional classroom) that is aligned to the intended outcomes, from the cognitive and skills domains to behavioural and attitudinal outcomes, of the geography curriculum.

References

Aldrich-Moodie, B., & Kwong, J. (1997). *Environmental education*. London: Institute of Economic Affairs.
Ballantyne, R. (1999). Teaching environmental concepts, attitudes and behaviour through geography education: Findings of an international survey. *International Research in Geographical and Environmental Education, 8*(1), 40–58.
Black, P. (2000). Research and the development of educational assessment. *Oxford Review of Education, 26,* 37–41.
Bracken, L., & Mawdsley, E. (2004). 'Muddy glee': Rounding out the picture of women and physical geography fieldwork. *Area, 36*(3), 280–286.
Chang, C. (2014). *Climate change education: Knowing, doing and being*. Abingdon: Routledge.
Cohen, P. (1995). Designing performance assessment tasks. *Education Update, 37*(6).
Cosgrove, D. (2008). Geography and vision: Seeing. *Imagining and Representing the World*. London & New York: IB Tauris.
Cotton, D. R. E. (2006). Implementing curriculum guidance on environmental education: The importance of teachers' beliefs. *Journal of Curriculum Studies, 38*(1), 67–83.
Cross, R. T. (1998). Teachers' views about what to do about sustainable development. *Environmental Education Research, 4*(1), 41–52.
Doyle, W., & Carter, K. (1984). Academic tasks in classrooms. *Curriculum Inquiry, 14*(2), 129–149.
Driver, F. (2001). *Geography militant: cultures of exploration and empire*. Oxford: Blackwell.
Fien, J. (1993). *Education for the environment: Critical curriculum theorising and environmental education*. Geelong: Deakin University Press. (Halstead, J. M & Taylor, M. J. (1996). *Values in education and education in values*. London: Falmer Press.).
Gipps, C. (1994). *Beyond testing: Towards a theory of educational assessment*. London, UK: Falmer Press.
Goodson, I. (1997). The changing curriculum: Studies in social construction (Vol. 18). Peter Lang Pub Incorporated.
Hagstrom, F. (2006). Formative learning and assessment. *Communication Disorders Quarterly, 28,* 24–36.
Ho, L. C., & Seow, T. (2015). Teaching controversial issues in geography: Climate change education in Singaporean Schools. *Theory & Research in Social Education, 43*(3), 314–344.
Huckle, J. (1985). Values education through geography: A radical critique. In D. Boardman (Ed.), *New directions in geographical education* (pp. 1–13). London: Falmer Press.
International Geographic Union - Comission on Geographical Education. (2016, August). International Charter on Geographical Education. Retrieved August 2016, from International Geographic Union - Comission on Geographical Education. Retrieved from http://www.igu-cge.org/Charters-pdf/2016/IGU_2016_def.pdf.
International Geographic Union - Commission on Geographical Education. (1992). International Charter on Geographical Education. Retrieved 2016, from International Geographic Union - Commission on Geographical Education. Retrieved from http://www.igu-cge.org/charter-translations/1.%20English.pdf

Jickling, B. (1992). Why I don't want my children to be educated for sustainable development. *Journal of Environmental Education, 23*(4), 5–8.
Krathwohl, D. (2002). A revision of Bloom's taxonomy. *Theory into Practice, 41*(4).
Lambert, D., & Balderstone, D. (2000). *Learning to teach geography in the secondary school—A companion to school experience*. London, UK: Routledge.
Lambert, D., & Hopkin, J. (2014). A possibilist analysis of the national curriculum in England. *International Research in Geographical and Environmental Education, 23*(1), 64–78.
McTighe, J., & Self, E. (2002). *Observable indicators of teaching or understanding*. TTL Academies.
Miller, J. L. (2005). *Sounds of silence breaking: Women, autobiography, curriculum* (Vol. 1). Peter Lang. Chicago.
Morgan, A. (2012). Morality and geography education. In G. Butt (Ed.), *Geography, education and the future* (pp. 187–205). London: Continuum.
Pinar, W. F., Reynolds, W. M., Slattery, P., & Taubman, P. M. (1995). Understanding curriculum as autobiographical/biographical text. *Understanding curriculum: An introduction to the study of historical and contemporary curriculum discourses,* 521–523.
Roberts, M. (2003). *Learning through enquiry: Making sense of geography in the key stage 3 classroom*. Sheffield: Geographical Assoiciation.
Rose, G. (2016). *Visual methodologies: An introduction to researching with visual materials*. London: Sage.
Sauer, C. O. (1956). The education of a geographer. *Annals of the Association of American Geographers, 46*(3), 287–299.
Sellan, R., Chong, K., & Tay, C. (2006). Assessment shifts in the singapore education system. Retrieved March 30, 2008, from Singapore Examinations and Assessment Branch http://www.iaea2006.seab.gov.sg
Seow, T. (2015). *Geographical inquiry through outdoor fieldwork*. Keynote address at the Geography Subject Chapter meeting, Academy of Singapore Teachers, 31 Aug 2015.
Singapore Examinations and Assessment Board. (2014). GCE O-Level Syllabus 2236. Retrieved 2016, from Singapore Examinations and Assessment Board: http://www.seab.gov.sg/content/syllabus/olevel/2016Syllabus/2236_2016.pdf.
Stoddart, D. R. (1986). *On geography and its history*. Blackwell.
Tomlins, B., & Froud, K. (1994). *Environmental education teaching approaches and students' attitudes*. Slough: National Foundation for Educational Research.
Tomlinson, C. (2014). *The differentiated classroom: Responding to the needs of all learners* (2nd ed.). Association for Supervision and Curriculum Development.
Tyler, R. W. (1949). *Basic principles of curriculum and instruction*. University of Chicago Press. Illinois, USA.
Voltz, D., Sims, M., & Nelson, B. (2010). *Connecting teachers, students and standards—strategies for success in diverse and inclusive classroom*. Alexandria, VA, USA: Association for Supervision and Curriculum Development.
Wiggins, G., & McTighe, J. (2005). *Understanding by design*. Alexandria, VA: Association for Supervision and Curriculum Development.
Williams, A. (2008). *The enemies of progress: The dangers of sustainability*. Exeter: Societas Imprint Academic.

Chew-Hung Chang is a geography educator serving as the co-chair of the International Geographical Union, Commission on Geographical Education, co-editor of the journal International Research in Geographical and Environmental Education, as well as the President of the Southeast Asian Geography Association. In addition to being a teacher educator, Chew-Hung has published extensively across areas in geography, climate change education, environmental and sustainability education.

Tricia Seow As an experienced geography educator, Tricia Seow is involved in the Fieldwork Exercise Task Force of the International Geography Olympiad and in the MOE Humanities Talent Development Programme. She also serves as the Hon-Gen Secretary of the Southeast Asia Geography Association. She enjoys fieldwork, and her research interests include teachers' knowledge and practice, field inquiry in geography and climate change education.

Part II
Teaching and Learning Geography Through Field Inquiry

Part II aims to enrich the discourse in teaching and learning geography through field inquiry with examples of fieldwork in Malaysia, Thailand and Singapore. The importance of field inquiry is critical especially in geography for students to understand the theoretical concepts in the classroom in practice. The examples include a field trip to the groynes at Bagan Tambang in Chap. 5, a water quality testing programme along Chao Phraya River in Chap. 6 and the learning journey programmes in Singapore in Chap. 7.

Field inquiry is an essential part of geography education to deepen students' theoretical concepts and their applications in the real world. Chapter 5 aims to explore the issue of geographic thinking and how the incorporation of well-selected learning resources for fieldwork can equip students with the skills to discover the world in geographical context. One of the key insights into this chapter is that changes in the real world might affect the field inquiry process; in the example given, changes in the sedimentation environment had changed the relevance of the groyne field as a geographical landscape to be studied. The author used the case of the changes to highlight the point that it is important to plan ahead for fieldwork so that the students can reap the most benefit out of the learning experience.

In Chap. 6, field inquiry takes the form of a water quality testing programme where students between the ages 13 and 15 years are identified and trained to test the water quality in Thailand where clean water is an increasingly scarce and important commodity. The chapter looks at the sustainability of this important resource with education for sustainable development (ESD) as an aim, taking a longer time perspective to consider the social and economic impacts of water quality issues. This field-oriented programme aims to allow students to learn through experiential inquiry and using this experience to improve their geographic thinking in the hope that they can solve similar real-world problems in the future, to benefit both themselves and their communities. It is through authentic and non-formal learning that we can have the biggest impact on geographical and environmental education.

Part II concludes with a case study about learning journeys in Singapore. Learning journeys are field-based learning experiences based on a trip or journey to a particular place, and this is especially important for geography where place is an important context to consider. Schools in Singapore conduct overseas learning experiences for students to experience foreign cultures, traditions and other social aspects of another society to broaden their horizons. Using a Strategize-Implement-Outcome (SIO) Model to examine the relative effectiveness of the learning journey, educators can identify if the specific educational outcomes or objectives have been met and can plan how to achieve desired results. The chapter also identifies some of the challenges associated with incorporating a learning journey as a field inquiry practice so to maximize the learning benefits and empower students to use geographic thinking from their experiences to solve real-world problems.

This part emphasized the importance of planning for fieldwork, conducting meaningful experiential learning and evaluating the specific outcomes of the fieldwork in learning geography outside the traditional classroom.

Chapter 5
What Happened to the Textbook Example of the Padang Benggali Groyne Field in Butterworth, Penang?

Tiong Sa Teh

Abstract Fieldwork in geography is considered to be "absolutely essential" as it "expresses a commitment to exploration and enquiry, and geography's concern to discover and to be curious about the world" (Lambert and Reiss 2014). While children may not have had the chance to see much of the world, one of the functions of textbook illustrations is to serve as a typical example through which teachers can take their students to the field for a closer study. This chapter explores the issue of geographic thinking, or its lack of, in shaping how we can select learning resources for fieldwork. The author uses a first-person narrative to develop an argument to explain the importance of using geographical thinking to consider the spatial and time scales, and the changing depositional environment over a few decades when selecting resources to support learning about a "real-world" case, through fieldwork. While textbooks may be published with new editions, the author argues the importance of studying the geographical context through geographical thinking when selecting resources to support teaching and learning beyond the traditional classroom, so that the curiosity to learn and discover about the world is set in the right context.

Introduction

When teaching about a place, it is difficult to do so without understanding the geographical context of the place. This is especially important when conducting fieldwork to teach geography. One of the functions of textbook illustrations is to serve as a typical example in which teachers can take their students to the field for a closer study. Some are natural features, while others are man-made. Large-scale natural features are more persistent than small-scale natural features or those man-made, which have a tendency to disappear when teachers are desperately searching for them with their impatient students. This reinforces the advice to

---

T. S. Teh (✉)
Singapore, Singapore
e-mail: tehtiongsa@gmail.com

© Springer Nature Singapore Pte Ltd. 2018
C.-H. Chang et al. (eds.), *Learning Geography Beyond the Traditional Classroom*,
https://doi.org/10.1007/978-981-10-8705-9_5

teachers to always do a reconnaissance session before taking students to the field. The Padang Benggali groynes in the Singapore Secondary 3 Geography textbook in the late 1997s were a good example of a groyne field in beach management. The groyne field and sandy beach were still there in 1996, but by the end of 2004 the coast had turned muddy and mangroves had encroached onto the muddy tidal flats. The sudden change in depositional environment from sand to mud is very unusual and deserves proper documentation to explain the phenomenon. Mud continued to accumulate since then, and the coast is now dominated by healthy mangroves. While mangrove coasts throughout Malaysia are being threatened by shoreline erosion, conversion and destruction, the Padang Benggali sandy coast has turned muddy, colonized by mangroves. The author will adopt a first-person voice to construct the critical narrative that documents the changing geography of the Padang Benggali coast.

My first visit to Padang Benggali in Butterworth was when I took my daughter on a countrywide coastal holiday in August 1992 and chanced to stop at the place. The whole sandy coast protected by a newly emplaced groyne field caught my attention, as the structure was unusually closely spaced and short (see Fig. 5.1). Later, I learned that this was because of the underlying soft sediments, which could not support longer, heavier structures.

Padang Benggali has a special place in my memory because I gave a slide of the short groynes to Ministry of Education, Singapore, and it was used in the Secondary 3 Geography textbook to illustrate different types of coastal structures (Curriculum Planning and Development Division, Singapore 1997). In 1995, my daughter used the textbook and she was amused to see the photograph and asked why I did not use another slide with her in it (Fig. 5.2).

Fig. 5.1 Padang Benggali sandy beach and short, closely spaced groynes (G2–G5) in August 1992

Fig. 5.2 Author's daughter in Padang Benggali August 1992. Nearest groyne is G2

Her classmates did not believe that I took the photograph and that she was just beside the groyne. Although different textbooks are used now, many teachers and students in Singapore are familiar with the Padang Benggali short groynes and may even be considering visiting the groynes when in Butterworth, as many teachers used that textbook when they were students. My advice is not to do it. The beach has completely disappeared, and most of the groynes are camouflaged by mangroves or stranded by land reclamation.

In November 1996, I took a group of Geography students from University of Malaya to the groynes at Bagan Tambang during their annual departmental fieldwork to Penang (Fig. 5.3). They mapped and measured the groynes and studied the impacts of the structure on littoral processes. The landscape had not changed much in the 4 years since I first visited Padang Benggali.

On the last day of 2004, when I was surveying the coast of Penang for a study on the impact of the December 2004 tsunami (Horton, et al. 2008), I revisited Padang Benggali after nearly 10 years. I could not recognize the place and could not believe my eyes when told by locals that it was indeed the place, which once had an attractive sandy beach and a groyne field. The coast had turned muddy, and mangroves had invaded the nearshore areas. A few of the old groynes were exposed by tsunami waves (see Fig. 5.4).

The presence of a wide belt of mangroves could have helped to lessen the impact of the tsunami. The changing sedimentation environment from sandy to muddy in such a short time is very interesting geomorphologically and needs to be documented. What happened in Padang Benggali may be a very rare phenomenon that requires a more detailed study later on. The following sections will provide the geographical context of the phenomenon and a discussion on its implication on using such textbook examples as learning resources for fieldwork preparation.

Fig. 5.3 Geography students from University of Malaya studying the groyne (G20) at Bagan Tambang at the southern end of Padang Benggali coast, November 1996

(a) Groyne exposed after December 2004 tsunami with wide mangrove belt in front

(b) Groyne field stranded inland, (G1-G4) 2004 December, with Sg. Abdul in background

Fig. 5.4 Padang Benggali groynes which were constructed to mitigate beach erosion now sit in an accreting mangrove environment

Background

The Padang Benggali coast had been classified as undergoing critical erosion in 1985, and coastal settlements had been threatened (Economic Planning Unit (EPU) 1985). The report suggested the construction of a groyne field as a suitable solution to mitigate shore erosion. A follow-up study was carried out by foreign consultants in collaboration with Department of Irrigation and Drainage (Eurosense Belfotop Consultant 1990). The study divided the whole Province Wellesley coast into subreaches with Padang Benggali occupying the northern section of Subreach 4 (Fig. 5.5). The study suggested that initial erosion at a rate of 4 m per year started between 1951 and 1977 at Padang Benggali and coincided with the construction of the Sg Muda Dam. Dredging of the access channel to the Butterworth Port may be a contributing factor (Kassim and Cheang 1991).

The coastal sediments consist of an upper sand layer (1–6 m), soft marine clay layer with shell fragments (6–12 m), middle sand layer (3–12 m) and a lower marine clay layer (10–12 m) and dense silty sand below. The subsoil is weak and highly compressible, and excessive settlement and slope instability are expected.

The original design was for the groynes to be spaced at 250 m interval with a minimum nourished beach of 30 m wide and a construction setback of 20 m. Coastline retreat was estimated at 120 m after 30 years under "do nothing" for Padang Benggali coast. The final recommendation from the study was that 11 rubble groynes were to be constructed in Robina and Royal Malaysian Air Force

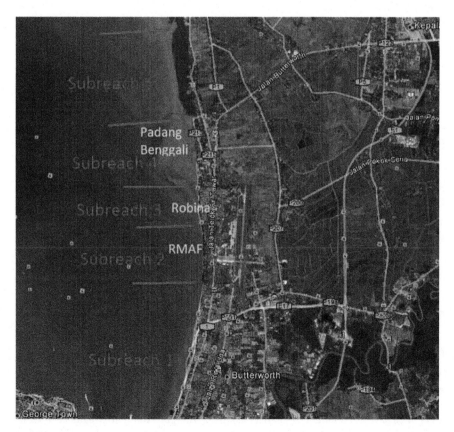

Fig. 5.5 Subreaches of Province Wellesley coast

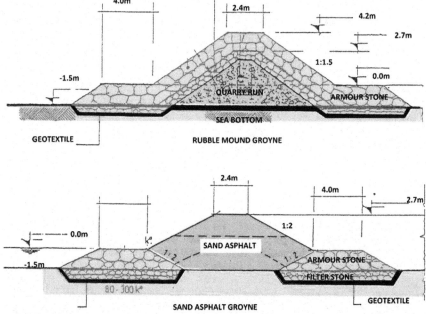

Fig. 5.6 Two of three groyne designs proposed for the Province Wellesley coast: the upper diagram of a rubble mound groyne was adopted. Groyne spacing of 150 m was for Robina. Recommended spacing for RMAF was 250 m

installation (RMAF) with a 30 m wide beach. Groyne spacing for Robina was to be 150 m and for RMAF 250 m. The recommendation for Padang Benggali coast was beach nourishment without construction of groynes. What eventually took place along the Padang Benggali coast was completely different from the recommendation. Several options in groyne design were proposed (Fig. 5.6), and finally rubble mound groynes were constructed.

Explaining the Geographical Context

The objective of this chapter is to document and explain the changing Padang Benggali coast from an eroding sandy coast protected by a groyne field to that of an accreting muddy mangrove coast, with a view to illustrate why geographical thinking is important when teaching and learning about a real-world context. The study area is the Padang Benggali coast from Sungai Abdul to Bagan Tambang bounded by Jalan Tuanku Putra (Fig. 5.7).

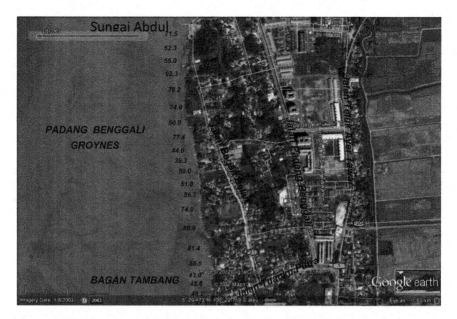

Fig. 5.7 Study coast from Sungai Abdul to Jalan Tuanku Putra: 2003 satellite image

Information on the changing Padang Benggali coast was obtained from field visits in 1992, 1996, 2004 and 2012 and from Google Earth satellite images for the year 2003, 2005 and 2010. The 2003 satellite image, the earliest available, was used to map the spacing and length of groynes. Additional information on the groyne field was from field mapping in 1996. Changes were described based on field observations and from information obtained from the above photographs and images. Satellite images were used to detect changes for the whole study coasts, and field photographs supplemented by satellite were used to detect changes for specific beaches/coasts.

Although it was recommended that the Padang Benggali coast was to be only nourished, a series of short groynes were constructed along the coast accompanied by beach nourishment. The 1.3 km long coast from Sungai Abdul to Bagan Tambang has a groyne field of 21 short groynes of varying length, constructed about 1991/92. The groynes are still clearly visible on the August 2003 Google Earth satellite image. As measured on the satellite image, the groynes are 43–86 m long and spaced at about 65 m. The length of the Padang Benggali groynes is much less than the recommended 150–250 m for the locations at Robina and the RMAF.

Changes for the Whole Study Coast from 2003 to 2010 Captured in Remote Sensing Images

The changes are illustrated and described in Fig. 5.8.

In January 2003, the former sandy coast had been replaced by a mud coast with young mangroves colonizing the mud flats. No beaches were observed on the satellite image. The groynes were still clearly visible, most with their seawards end in the sea and landwards end among mangroves (G14–G21); a few were completely stranded inland with mangroves separating them from the sea (G6, G7), and others had newly colonizing mangroves in front (GG3, G4, G5).

Groyne 1 which was part of the seawall that lined the coast near Sungai Abdul served as a training wall and was largely unchanged since emplacement. By April 2005 about 4 months after the tsunami, a beach (probably shelly) had developed at the outer edge of the mangroves between G1 and G4, probably associated with the effects of the tsunami waves. Mangroves fully dominated the coast between G5 and G8, and the groynes were overtopped by mangroves. The coast to the south remains largely the same. Five years later in February 2010, mangroves had extended seawards between G1 and G4 and the groynes became stranded behind mangroves. A sandy beach developed between G4 and G5 as well as in short stretches to the south. Mangrove encroachment into the sea and small reclamation works resulted in nearly all the groynes being stranded inland along the Padang Benggali coast.

The northern part of the coast had advanced as a result of natural accretion of fine sediments, and this caused the groynes to be completely stranded inland by 2010. However, the southern part of the coast advanced because of natural accretion and land reclamation. Land reclamation may be "opportunist" type to convert mangrove areas into land. The series of coastline change in the south between 2003 and 2010 is shown in Fig. 5.9.

(a) 2003 January Google earth (b) 2005 April Google earth (c) 2010 Google earth

Fig. 5.8 Coastline change Padang Benggali coast 2003–2010 and impacts on groynes

Fig. 5.9 Bagan Tambang groynes (G15 to G21) stranded inland by initial sedimentation (2003–2006) followed by land reclamation (by 2010)

Changes Captured on the Ground from 1992 to 2012

From 1992 to 1996, the whole Padang Benggali coast of groynes and a nourished beach remained essentially the same (Fig. 5.10). The coast and fishing village behind the beach had been protected by the emplacement of groynes around 1991.

During the 2004 December visit to the Sungai Abdul end of the beach, the sandy beach had disappeared and most of the groynes were stranded far behind a fringing belt of mangroves. The tsunami waves had uprooted some of the young mangroves and exposed some of the groynes (Fig. 5.4). The groynes photographed during the tsunami study are G1 to G4.

During a subsequent visit to the study coast in September 2012, there appeared to have been some shoreline retreat and that some of the groynes previously stranded had come to front the sea (Fig. 5.11). Groyne 1 which served as a river mouth training wall was clearly recognizable, but groyne 2 stayed stranded and hidden below coastal shrubs. Groyne 3 had its distal end adjoining the sea after a new phase of shoreline erosion eroded away the land in front of the groyne. The coast where groyne 11 is located also appeared to be undergoing erosion.

(a) 1992 Sungai Abdul end of beach nourishment and groynes

(b) 1992 Sungai Abdul end, with fishing village behind beach

(c) 1996 Groyne field at Bagan Tambang

Fig. 5.10 Beaches and groynes along the whole Padang Benggali coast 1992–1996

(a) Groyne 1 beside Sg. Abdul 2012 September

(b) Groyne 2 hidden under coastal bushes-2012 September

(c) Groyne 4 fronting the sea 2012 September

(d) Groyne 11 beginning to front the sea after shoreline retreat-2012 September

Fig. 5.11 Condition of groynes near Sungai Abdul end in September 2012

Discussion

The study shows that what was eventually implemented along Padang Benggali coast in the early 1990s to mitigate shore erosion did not follow the recommendation of 1990 consultancy report. At the time of the study, it was pointed out that several uncertainties remained and future development had been planned for the Butterworth coast. These included the North Butterworth Container Terminal (NBCT) as well as the proposed coastal expressway, in which the impact on the Butterworth coast was largely unknown (Fig. 5.12). In addition, there was already a proposed plan to reclaim the whole Butterworth coast (Shamsul and Teh 1991).

The recommended mitigation measure for Padang Benggali coast was nourishment without construction of any coastal structures. Instead what actually took place in 1992 soon after was an emplacement of a groyne field with beach nourishment. The reasons for this are unknown to the public. The groyne fields remained intact up to 1996. But sometime between 1996 and January 2003, the coast turned muddy and mangroves started colonizing the flats. By December 2004, the groynes were stranded behind a wide mangrove belt. The sudden change in sedimentation environment from sand to mud was unexpected and needs explanation. At the time when the groynes were emplaced, major changes to the Butterworth coast were taking place. Land was being reclaimed for the NBCT accompanied by massive dredging for the approach channel to the Port. The project

Fig. 5.12 Construction of North Butterworth Container Terminal and Coastal Expressway resulted in impacts on coastal processes along Butterworth coast. Map source Google Earth 2010

was completed in 1994. The impact on littoral drift along Butterworth coast was recognized in a 1990 study, but the dredging and disposal of the spoils may have escaped attention. Proper disposal of the dredged spoils would not result in adverse impacts. But if the spoils were improperly disposed and washed back to the coast, then this might explain why the sandy coast eventually turned muddy. This was the explanation given by locals when asked about the changing coast. This is their perception on the reason Padang Benggali coast turned muddy and became mangrove fringed. Satellite images also suggest that after accretion of mangroves at least the southern Padang Benggali coast was reclaimed and unprotected. There appears to be a new phase of erosion along the whole coast, and shoreline retreat is beginning to expose the distal end of groynes. The big question is "Will shoreline erosion revert the coast back to a sandy beach with a groyne field?"

Another issue that needs to be addressed is why groynes were emplaced along a coast that was obviously going to be modified by development and land reclamation. Was the decision to construct a groyne field too hasty in view of the impending coastal modification? Would a less costly beach nourishment programme suffice or even a "do nothing" option? The groyne field became completely redundant about 10 years after construction. It is also surprising that the coast where groynes and beach nourishment were recommended ended up being defended by an extensive rock revetment. Fortunately, it turned out to be the right decision as the armourment installed fortuitously helped to protect the coast from the 2004 tsunami (Fig. 5.13).

Fig. 5.13 Massive rock bund along Robina coast helped to mitigate the 2004 tsunami effects. Photograph taken five days after tsunami on 3 December 2004

(a) Earliest rock groyne in Penang (b) River groynes at Pekan

(c) Dungun groynes of concrete spine and tetrapods (d) Groynes and river mouth training wall of North Melaka

Fig. 5.14 Selected groyne field s in Peninsular Malaysia

The use of groynes to mitigate erosion appears to have lost favour in Penang. No new groynes had been constructed after the Padang Benggali groynes. The earliest rock groynes behind Dalat School constructed by Australian soldiers remain the only groynes on the island. Elsewhere in Peninsular Malaysia, there was a series of groyne construction in the 1980s and 1990s (Fig. 5.14) but construction also appeared to have slowed down (Teh and Yap 2003).

A useful lesson learned is that proper dumping of marine mud and silt can be used to mitigate an eroding mangrove coast, in which the waves will help to spread the sediments and newly accreted areas can be colonized naturally by mangroves without any artificial mangrove replanting.

Conclusion

The emplacement of a groyne field along the Padang Benggali coast in the early 1990s in response to shoreline erosion may have been too hasty. A "do nothing" option would have been better. Improper disposal of dredged wastes related to the development of the North Butterworth Container Terminal probably resulted in turning the former sandy coast to a muddy coast. Mangrove encroachment onto the mud flats and minor land reclamation stranded the groynes inland. This sudden change in sedimentation environment is unusual. With the change in sedimentation environment, the textbook example of a groyne field and its functions lost its relevance.

The question set out in the title of this chapter "What happened to the textbook example of the Padang Benggali groyne field in Butterworth, Penang?" allowed us to think through the geographical reasons for the changed physical landscape. The empirical research conducted by academic geographers has highlighted the need for teachers to understand that coasts are subjected to changes at spatial and temporal scales. While natural hazards may change a physical landscape radically (as in the case of the 2004 tsunami), development factors such as those highlighted above would have also changed the landscape. The point that is argued here is not that changes are inevitable, but rather geographical thinking will allow the teacher to ask questions about how the landscape could have changed. Some of the tools employed in this study are easily accessible, such as Google Earth. Perhaps teachers can use these tools to conduct preliminary research before going out to the field, thinking that they will see the perfect textbook example.

What happened to the textbook example along Padang Benggali applies to other textbook examples. Their accuracy, existence and relevance must be constantly checked. Using local or regional examples you are familiar with and can be easily checked from time to time would be preferred. You do not want to be caught in a situation where the coastal features you expect do not exist on the day you conduct your field study.

References

Curriculum Planning and Development Division, Singapore. (1997). *Understanding geography* (Vol. 3). Singapore: Longman Singapore Publishers (Pte) Limited.

Eurosense Belfotop Consultant. (1990). *Final review report on Coastal Defence Works*. Butterworth, Malaysia: Economic Planning Unit.

Economic Planning Unit (EPU). (1985). *National coastal erosion study*. Butterworth, Malaysia: Prime Minister's Office.

Horton, B., Bird, M. I., Cowie, S., Ong, J. E., Hawkes, A., Tan, S., et al. (2008). ndian Ocean tsunamis: environmental and socio-economic impacts in Penang. *Malaysia. Singapore Journal Tropical Geography, 29*(3), 307–324.

Kassim, M., & Cheang, T. (1991). Coastal Engineering in coastal development. In *Proceedings, IEM/ICE Joint Conference* (pp. H1–H20). Kuala Lumpur.

Shamsul, B., & Teh, T. (1991). Coastal land reclamation and future sea level rise implications in Malaysia. *Malaysian Journal of Tropical Geography, 22*(2), 145–162.

Teh, T., & Yap, H. (2003). Penang. In E. Bird, *The World's Coast* (Vol. Online). Kluwer Academic Press.

Tiong Sa Teh is a coastal zone management specialist and has vast experience in the field of coastal geomorphology, climate change and sea level rise, application of satellite remote sensing and GIS. He was actively involved in assessing the impact of sea level rise and combined extreme events for shore management studies. He also was actively involved and contributed to the Intergovernmental Panel for Climate Change—National Communication Report.

Chapter 6
The River Guardian Program for Junior High Schools on the "River of Kings," Thailand

Supitcha Kiatprajak and Lynda Rolph

Abstract A country's main river is like the central bloodline sustaining the majority of people who live there. The Chao Phraya River is Thailand's main river, running through Ayutthaya, Thonburi, and Rattanakosin or Bangkok, three important capital cities from the days of Siam to present-day Thailand. Over time, our relationship with this river has changed and our actions have degraded it. Pollution, canal building, and damming have contributed to ecosystem changes. The best way to try to conserve our main river is to enable new generations to learn about and to love their own resource. With this desire in mind, the River Guardians Project was created. The River Guardians Project is one of the programs administered by Traidhos Three Generation Barge Program (http://barge.threegeneration.org/), working in the field of education for sustainability at different locations in Thailand. A group of five Thailand government junior high schools (M1-M3 level, or 13–15-year-olds) in Bangkok were identified and trained to test the water quality in their section of the river, going from near the city boundary, downstream, to the heart of the city. Dissolved oxygen (DO), biochemical oxygen demand (BOD5), E. coli, nitrate (N), phosphate (P), pH, water temperature, total dissolved solids (TDS), and turbidity were analyzed as representative parameters for the quality of the river in this research. Trends in water quality were observed particularly in relation to local land use patterns. Although coordination with the schools at times can be challenging, overall the teachers felt that the students have benefited from the experience academically and it has given them an appreciation for the connection of water and community. The Education for Sustainable Development philosophy behind the program, the logistics of creating the program, water quality testing results, and lessons learned are presented in this chapter.

S. Kiatprajak (✉) · L. Rolph (✉)
Traidhos Three-Generation Community for Learning, Chiang Mai, Thailand
e-mail: backtotheorigin@gmail.com

L. Rolph
e-mail: lynda@compasseducation.org

Introduction

Water is the most important element on this planet earth, and no one can live without it. Yet, availability of clean water varies tremendously from country to country. Pruss-Ustun et al. (2014) concluded that 842,000 diarrheal deaths occurred globally in low- and middle-income countries in 2012 as the result of inadequate water, sanitation, and hand hygiene. Asian rivers are among the most polluted in the world, with three times as many bacteria from human waste as the global average (United Nations University 2016). Water consumption has almost doubled in the last 50 years and globally; the acreage equipped for irrigation increased from 193 to 277.1 million hectares between 1980 and 2003; the largest proportion of this irrigated land is in Asia (Food and Agriculture Organisation of the United Nations (FAO) 2016).

If we are not concerned about our water resource, we may experience considerable social and economic stress in the near future. Therefore, the River Guardian Program was set up to help the new generation learn about and love their water resource.

The Chao Phraya River is Thailand's main river that provides nourishment to its people in both direct and indirect ways (Fig. 6.1). With its low alluvial plain forming the central landmass of the country, it runs through Ayutthaya, Thonburi, and Rattanakosin (or Bangkok), three important capital cities from the historic times of Siam to present-day Thailand, before it empties into the Gulf of Thailand. The Chao Phraya Watershed covers about 30% of Thailand's land area or about 160,000 km^2 (Komori et al. 2012). The flow on the Chao Phraya River follows the seasonal monsoon pattern, with low flows occurring in the dry season, December to May, and with the flow rising to peak around October (Fig. 6.2).

Water is an extremely important element in the Thai peoples' lives because they have lived with and on the water since historic times. It provided the main form of transportation through rivers and canals, when roads were absent in the early periods of their history. This is the reason why many important places such as the Grand Palace, temples, government offices, mosques, and also residences were settled along the banks of the river as well as along the canals (Sitthithanyakij 2007). Daily life is connected to the river or the water also because Thailand is an agricultural country with extensive rice fields and fruit gardens. Rice fields always employ small canals or the water wheel to pump the water into their rice fields (Tiptus 2000). Moreover, there are many other careers such as fishing and commercial navigation that are supported by the water from the river. In addition, most of the bricks that were used to build the Grand Palaces in Sukothai, Ayutthaya, Thonburi, and Rattanakosin were created by the clay taken from the Chao Phraya River or its tributaries.

During the rainy season, flooding frequently occurs in the middle part of Thailand and this brings nutrients to the farmers' fields (Pollution Control Department 2003). The local people also learned how to live with this situation as we see from the architectural styles of the houses that have high stilts and a very

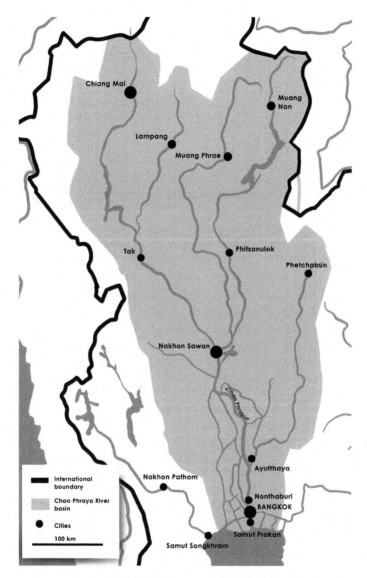

Fig. 6.1 Chao Phraya Watershed (from the Working Group of the Office of Natural Water Resources Committee of Thailand) (*Source* adapted from UNESCO 2006, p. 391)

sharp roof. Thai houses traditionally were designed to use the space under the house for activities such as cooking and family gatherings and to be safe in the flooding season (N Paknam 1988). And yet, Thai traditions are changing. Some have argued that Thai people are losing their connection with the water, particularly in the urban and peri-urban areas of Bangkok (Suwanarit 2012).

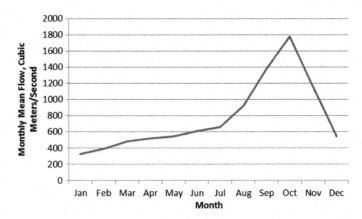

Fig. 6.2 Monthly mean flow, 1976–1984, for the Chao Phraya River at Nakhon Sawan. *Source* Oakridge National Laboratory, Global River Discharge Database, RivDIS Project (*Source* (National Aeronautics and Space Administration (NASA) 2016)

The Chao Phraya River has been degraded through anthropogenic activities (e.g., Patarasiriwong 2000; Ongsakul and Sajor 2006; Price et al. 2012). Since the Ayutthaya era, international trade has been welcomed in Thailand (Reid 1990; Villiers 1999), the Chao Phraya has been dredged, and canals were constructed to straighten the meandering river course for the purposes of navigation. However, all is not gloom and doom. From the water quality reports of many organizations including those by the River Guardian Project, water from the upper and middle part of Chao Phraya River is still of an acceptable quality. It is only the lower part of the river where water quality is quite poor (see also Simachaya 2003). If we are not aware of the value and the importance of the river, we run the risk of severe environmental impairment for future generations.

The best way to try to conserve our main river is to enable new generations to learn about and to love their own resources. Once they feel a connection to the river and realize its value, we should build their capacity so that they can help each other to take good care of their own resource. The aim of this chapter is to describe a water testing program that was established with five Thai government schools along the Chao Phraya River and present results of the testing and discuss the implications for teaching and learning. To provide context and a framework for the water quality testing work, we will first describe the River Guardian Program through a case study approach, showing that non-formal, field-oriented education can provide authentic research experiences that solve real-world problems, benefitting both the students and the community.

Background of Traidhos Three Generation Barge Program

The Traidhos Three Generation Barge Program has been running since 1995, when Mom Luang Tridhosyuth Devakul converted a teak rice barge into a floating classroom on the Chao Phraya River (Fig. 6.3). The program was developed to showcase experiential environmental education, based on examples in Canada and the USA. The early days of the program were limited to working on the Chao Phraya River with international and Thai government and private schools and offering training to Thai teachers wanting to implement the Thai Ministry of Education's child-centered curriculum and developing critical thinking initiatives. The program was subsequently extended into a Watershed Program, with the addition of activities in mountain areas, rainforest habitats, and at marine sites of the watershed, enabling us to facilitate students with a consistent theme but different watershed contents over a number of years. Fieldwork provides a number of benefits through hands-on practice that often gives more meaning to theoretical material taught in the classroom, which can positively influence cognitive processes and affective learning (Kern and Carpenter 1984; Smith 1999; McGuinness and Simm 2005; Boyle et al. 2007; Dummer et al. 2008; Brundiers et al. 2010). The authors have found the fieldwork programs have been well-received by Thai schools and students and international schools and students alike.

The UNESCO Decade of Education for Sustainability prompted us to re-examine the work we were doing and to realign our program with those ideals. Our Environmental Education (EE) programs have already contributed to Education for Sustainable Development (ESD) but we became more aware of the process of systems thinking and the skills of envisioning, critical thinking, and the importance of networking (discussed in more detail in the next section). With the curriculum enriched by new processes and ways of thinking, we wanted students participating in our programs to be exposed to skills relevant to twenty-first-century thinking.

The River Guardian Program is a systems-thinking approach to education for sustainability. The water quality testing task has been framed in the geographical context of both the Chao Phraya Watershed and the immediate environment around

Fig. 6.3 Three Generation Barge (left) and class instruction on the barge (right)

the various schools. The partnership formed between Buffalo State, State University of New York, and the Traidhos Three Generation Barge reminds us of the global nature of today's world dependent on the same natural resources regardless of wealth, status, or location.

The River Guardian Project discussed in this chapter was conducted in four phases over two years with five Thai government schools:

Phase 1. Watershed awareness: Games and activities were developed for students to appreciate that they have a watershed address and that the Chao Phraya River near their school comes from somewhere and goes to somewhere. This is to raise awareness that the river is a system.

Phase 2. Water quality testing and data collection: Introduction to water quality test kits, introduction to what the tests mean, introduction on how to record and manage data, and testing and river bank observations.

Phase 3. Community investigation around school area, using the AtKisson Compass of Sustainability (Steele 2011). Systems-thinking approach was used to identify what is happening in the immediate area that may be impacting on the water quality (water testing also continued).

Phase 4. Student led community action to address an issue identified in phase 3. Presentations by all River Guardian schools and the submission of their final reports.

Education for Sustainable Development (ESD)

Over the last twenty years, there have been many definitions of "sustainable development." Perhaps the most famous and long-lasting is the definition offered by the Brundtland Report, where sustainable development is that which meets the "needs of the present without compromising the ability of future generations to meet their own needs" (World Commission on Environment and Development 1987, p. 7). Indeed, ESD falls within the larger umbrella of environmental education and the chapter will consider these two terms to refer to the same goal of enabling our learners to care for and take action on their environmental future.

In our discussion, however, sustainable development will be taken as continuous change in the direction of sustainability as defined by the AtKisson Group. Education for Sustainable Development is any learning that creates change and that leads in the direction of sustainability (Steele 2011). By the end of the UNESCO Decade of Education for Sustainable Development in 2014, much has been written about ESD. As with the concept of sustainable development itself, different groups and countries have different (at times conflicting) views and approaches to ESD (e.g., Jickling 1994; de Haan 2006; Vare and Scott 2007; Venkataraman 2009; Mogensen and Schnack 2010; Wals and Keift 2010). The value of both formal and non-formal educational approaches for ESD has been highlighted by a number of researchers (e.g., Haigh 2006; Tilbury and Wortman 2008; Brundiers et al. 2010; Wals and Keift 2010), and the River Guardian Project is a good example of how non-formal education and fieldwork can enhance the student learning experience.

UNESCO Clearly Defines ESD

Education for Sustainable Development aims to help people to develop attitudes, skills, and knowledge to make informed decisions for the benefit of themselves and others, now and in the future, and to act upon these decisions (UNESCO 2011).

So how does ESD as a teaching tool enrich the River Guardians Project? Essential to ESD are the following skills identified by Tilbury and Wortman (2008) and Tilbury (2011):

1. Envisioning
2. Critical thinking and reflection
3. Systemic thinking
4. Building partnerships
5. Participation in decision-making

Envisioning

We want the students involved in the River Guardian Project to be able to imagine a pollution-free Chao Phraya River. We want them to understand how the river that passes their school is part of a system. To this end, the first few sessions with each school did not focus on water testing but instead the students were engaged in a program of general environmental awareness, their watershed address, and understanding that the river comes from somewhere and flows on to somewhere else, hence connecting their school and community to the wider watershed. If students are to be motivated to make a difference, they need to envision a clean and sustainable river. This will help them to identify the goal for part two of the project where children have to imagine a world in which people from all backgrounds and levels of expertise are engaged in a process of learning for improving quality of life for all within their community and beyond for future generations, in a world where people recognize what is of value to sustain and maintain, and what needs to change through "reflecting, understanding, asking, making choices, and participating in change for a better world" (Tilbury 2011). Envisioning identifies relevance and meaning for our students, it explores how change can be achieved, as it offers direction and energy to take action, and it results in the ownership of visions, processes, and outcomes.

Critical Thinking and Reflection

Critical thinking and reflection challenge our current belief system and the assumptions underlying our knowledge, perspectives, and opinions about how Thai children learn. This is very important in Thai government schools since students are traditionally taught to respect the opinion of the teacher and are not encouraged to question material that is presented.

Critical thinking helps River Guardian students to reflect and develop the ability to participate in change, as it provides a new perspective and promotes alternative ways of thinking. In phase two of the project, students reviewed what was happening around their school areas from multiple viewpoints—environment, economy, society, culture, and personal well-being—since these are the strands that contribute to a sustainable system. By doing so, a holistic view of all that is happening in their area was created and this gave confidence to students to address the unsustainable practices which may be contributing to the poor water quality in their section of the river.

Systems Thinking

Systems thinking is "a unique perspective on reality—a perspective that sharpens our awareness of the whole and how the parts within those wholes interrelate" (Waters Foundation 2016). One might ask why systems thinking is important for learners. Exploration of dynamic complexity is a highly motivating learning experience for students. Their learning is enhanced by the "real" nature of the problems that they explore…[and] it creates tremendous potential for engaging students in powerful learning experiences (Waters Foundation 2016). Our River Guardian students move from a general watershed understanding to specific knowledge of the health of the river at their school and then in phases three and four, and get the chance to connect their testing results to what is really happening around their school. Students will start to make real-life connections as they use the AtKisson Compass of Sustainability, and that once they see the river as part of a system, they will be able to suggest where changes can be made, making the system more sustainable.

Systems thinking allows students to develop a number of higher order skills. We start to seek the big picture, look for change over time, and become aware of delays in change. By looking from a number of perspectives, students learn to understand an issue more fully. Systems enable the learner to consider cause and effect and where change can be innovated; systems allow us to understand the effect our actions will have more easily.

Imagine a world where decision makers "saw the whole picture" and can honor the connections between their actions and local, regional and global issues; a world where people and communities have the skills to understand links between thinking, actions, and impact across the world, and where they are empowered to address core problems and not just the symptoms (Tilbury 2011).

Participation in Decision-Making Empowers Oneself and Others

Putting decision-making and responsibility for the action in the hands of the participants creates a sense of ownership and commitment to action for the River Guardian participants which will help build capacity for self-reliance and self-organization and empower individuals to take action.

By rooting River Guardians in the theory of Education for Sustainable Development, the program is characterized by being interdisciplinary, values-driven, and a shared learning between teachers, learners, and our partners. It is locally relevant and seeks innovation. It enables students to stop looking at fragments of a situation and to look at connections to other things.

The four phases of River Guardian Project, watershed understanding, water testing, community investigation, and action project, will not only achieve a class water testing project but it will build the capacity of each participant to understand their locality and to develop thinking skills, knowledge, and understanding that they can apply to many aspects of their life.

Water Quality Testing Program

Target Group

With the desire to provide a dynamic educational experience in sustainability that is rooted in the local issue of water resources, five Thailand government junior high schools (M1–M3, 13–15-year-olds) were identified and trained to test the water quality in their section of the river using the water testing kits that were supported by Buffalo State, State University of New York. The project fits with the school's curriculum and study development program as the students have to do the water testing and learn about basic environmental chemistry in the science subjects, and this will help them, for example, in developing a "junior project" in the following year, which is also part of phase four of the River Guardian Program. The schools are located in Pathum Thani, Nonthaburi, and Bangkok, and three of these were selected to represent conditions along the lower Chao Phraya River. The fourth school was selected to represent the former Chao Phraya River which is now known as Bangkoknoi Canal. The fifth school was selected to represent the boundary conditions upstream of the Greater Bangkok area (Fig. 6.4).

Fig. 6.4 Location of the River Guardian Schools on the Chao Phraya River. The pin furthest south on the Chao Phraya River represents the office location of the Traidhos Three Generation Barge Program

Kanarajbumroong Patumthani School: This school is one of the science centers among other Thai schools. The school is located at the upstream of the Chao Phraya River in Patumthani Province and opposite Patumthani fresh market which is still in the perimeter near Bangkok.

Sriboonyanon School: This school is located downstream from the first school in Nonthaburi Province. They are a very active school that has a record of success with science projects.

Suksasongkror Bangkruay School: This school is located in the Bangkruay district, Nonthaburi Province. Their location is the border between Nonthaburi Province and Bangkok. This school was chosen because all of the students are members of hill tribes from the northern part of Thailand, which is the source of the Chao Phraya River.

Wimuttiyaram Wittayakom School: This school is located in Bangkruay district as well but at the end of the Chao Phraya River, downstream of the Bangkruay Power Plant.

Dipangkornwitthayapat (Wat Noi Nai) School: This school is located on Bangkoknoi Canal which is used to be the former Chao Phraya River before the present river was dug up.

Procedure

Each school collected grab samples from the bank of the Chao Phraya River in front or nearby their school using a bucket. Sample collection was done on the first day of every month. The water was then analyzed using testing kits provided by Buffalo State, State University of New York, and included nine tests:

1. E. coli—using the Coliscan Easygel system (www.micrologylabs.com/). Each test comes as a self-contained unit that includes the growth media in a disposable plastic vial, a specially treated plastic, disposable petri dish, and a disposable pipette. Normally, 1 mL of water is extracted from the sample and dispersed into the growth media vial. The vial is then gently swirled to fully mix the water and growth media and poured into the petri dish. The proprietary coating on the petri dish produces a chemical reaction with the growth media and bacteria in the water sample such that E. coli colonies turn blue or purple, whereas the total coliform colonies are pink. In this study, only the E. coli colonies were counted. Colonies are counted after 48 h, and an advantage of the system is that no specialized incubation equipment is needed. Incubation is done at room temperature, and the analysis can be completed on a laboratory bench. Irvine et al. (2011) compared the Coliscan results to E. coli levels determined from a split sample using the standard membrane filtration method at a New York State-certified laboratory and showed the Coliscan results were similar over a wide range of E. coli levels.
2. pH—Extech Instruments PH60 waterproof pH/temperature pen.
3. Temperature (difference between two sites)—Extech Instruments PH60 waterproof pH/temperature pen.
4. Turbidity—Secchi Disk. The Secchi depth can be converted to the NTU or JTU scale using the conversion graph provided by Mitchell and Stapp (1995).
5. Total Solids—evaporation method.
6. Dissolved Oxygen (DO) and BOD5—CHEMetrics Oxygen (dissolved) kit (http://www.chemetrics.com/). This kit uses a colorimetric approach employing the indigo carmine method (Gilbert et al. 1982). A 25 mL sample is used for analysis. A separate 25 mL sample was collected at the same time and stored in an amber HDPE bottle. The same kit was used to analyze DO in the amber bottle 5 days later.
7. Nitrate—CHEMetrics kit which uses a colorimetric approach employing the cadmium reduction method (APHA Standard Methods, 21st ed., Method 4500-NO3- E (APHA 2005)). A 15 mL sample is used for analysis.
8. Phosphate—CHEMetrics (reactive ortho) kit which uses a colorimetric approach employing the stannous chloride method (APHA Standard Methods, 21st ed., Method 4500-P D (APHA 2005)).

These parameters were selected to be consistent with the National Sanitation Foundation (NSF) Water Quality Index (WQI) (Brown et al. 1970), and the kits have been calibrated in New York state (Stephen Vermette, Professor, Buffalo State, State University of New York, pers. Comm. and in Singapore (Lok 2014; Ng 2014) to provide robust results. These kits are easy to use and are cost-effective.

The WQI can be a useful tool for summarizing and communicating complex water quality information to the public and has been applied in many countries worldwide (Bhargava 1983; House and Ellis 1987; Dojlido et al. 1994; Palupi et al. 1995; Wills and Irvine 1996; Pesce et al. 2000; Bordalo et al. 2001; Cude 2001). The students use the data to calculate the WQI, which synthesizes the results of all water quality tests into a single value between 0 and 100 (closer to 100 being better water quality). The WQI facilitates interpretation of the data and also provides the students with experience in graphing data and making simple calculations. The calculations for the WQI can be done using a Microsoft Excel spreadsheet developed by Kim. N. Irvine; Buffalo State, State University of New York or manually, using the water quality (Q-Value) rating curves and tables provided by Mitchell and Stapp (1995) (Figs. 6.5 and 6.6).

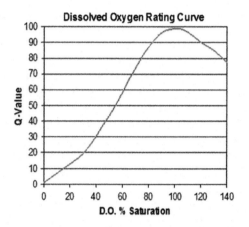

Fig. 6.5 Example of Q-value rating curve, in this case for dissolved oxygen (based on Mitchell and Stapp 1995)

Analyte	Test Result	Q-Value	Weighting Factor	Total
DO			0.17	
E. Coli			0.16	
pH			0.11	
BOD			0.11	
Temperature			0.1	
Phosphorus			0.1	
Nitrate			0.1	
Turbidity			0.08	
Total Solids			0.07	

Overall Water Quality Index Score: _____

Fig. 6.6 An example of the tabulation to facilitate manual calculation of the WQI

Results and Discussion for the WQI

The sampling process (see Fig. 6.7) started in February 2012; however, some schools started in June 2012 because the historic flood of 2011 delayed the program. Results of the sampling in terms of the WQI are summarized in Fig. 6.8, and generally the WQI values fall into the "medium" to "good" range following the scale proposed by Mitchell and Stapp (1995). Results were somewhat unevenly reported, and this is discussed in the next section in more detail. The Wat Noi Nai School continued the project past November 2012 and into September 2014, and the full set of results for this school are shown in Fig. 6.9. The results do not show any real seasonal variation, as we had expected in planning the project. In fact, human factors seem most likely to dominate the water quality characteristics of the area. For example, even though the Kanarajbumroong Patumthani School is furthest upstream and above the Bangkok CBD, it is located opposite to the Pathumthani market and throughout the day ferries run back and forth, continually pouring exhaust and oil discharges into the water. The area is residential with septic tanks frequently emptying into the river, while market vendors tip waste into the water. Wat Noi Nai School is located in the small klong, Bangkoknoi. It is an area of

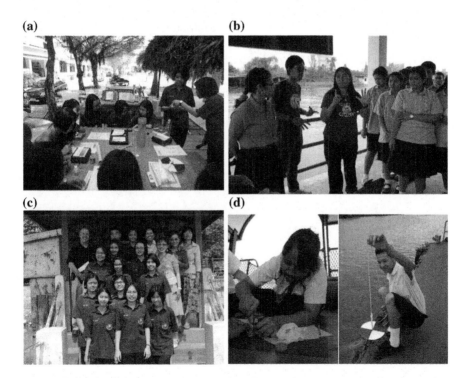

Fig. 6.7 Photographs of students on the project. **a** Students at Kanarajbumroong Patumthani School conducting water tests. **b** Supitcha Kiatprajak with colleague explaining the water testing kits to Dipangkornwitthayapat (Wat Noi Nai) School. **c** Students, teachers, and Buffalo State partners at Kanarajbumroong Patumthani School. **d** Suksasongkror Bangkruay School and Dipangkornwitthayapat (Wat Noi Nai) School student testing water

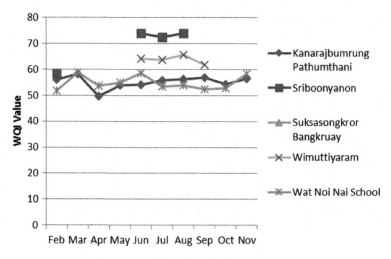

Fig. 6.8 WQI results for all reporting schools, February–November 2012

mostly gardens and orchards with some redeveloped residential units. It also is a popular site for tourist long-tail boats. When the water level is low, these boats stir up the sediment on the bottom of the canal, contributing to the volume of black, turbid water.

The results from the WQI have allowed the students to think deeply about a few things. For one, they had to correlate the findings to understanding the spatial location of their sites. In addition, the students also had to explain how the situation at the site will help explain the results. The example of Kanarajbumroong Patumthani School is worth mentioning here, where students are provided the

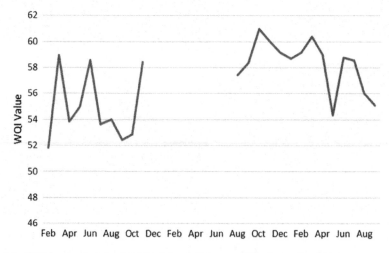

Fig. 6.9 WQI results for Wat Noi Nai School, February 2012–September 2014

opportunity to link their knowledge about turbidity with the real-world context of sediments stirred up by passing boats. However, this is just the WQI portion of the program and there are other lessons learnt throughout the entire project.

Lessons Learned and Concluding Thoughts

Overall, we believe we delivered a successful program (e.g., Figure 6.10) but we also identified shortcomings that need to be considered, as well as some factors that impacted the project that were beyond the authors' control. The teachers at the Kanarajbumroong Patumthani School said that all of students who joined the project gained valuable experience through their participation in the program and it helped develop their leadership skills. The students felt confident to work with the water testing kits and started to be able to think and to organize a junior project of their own. Although the junior project did not use the WQI directly, the data that they gathered from this project suggested that the quality of the Chao Phraya River is variable over one year and this inspired the students to think about possible projects. Subsequently, the junior project came about after testing their school sewage water. They created a filter after testing different kinds of filter material, including jackfruit peel charcoal, mangosteen peel charcoal, orange peel charcoal, and coconut peel charcoal. The conclusion of the project was that the mangosteen peel charcoal was the best to filter the water. This school had an excellent experience, in part because of the teacher leadership and in part because the school has a strong science/research tradition.

Wat Noi Nai is another school that worked very enthusiastically on the project but after November 2012, they could not conduct the testing for a period of time because their water testing box was stolen from the school laboratory. We replaced the kits for them at the middle of the year 2013, and they started to collect the data again in August 2013 and finished their data collection in September 2014. Unfortunately, the lead teacher moved to another school.

The Sriboonyanon School submitted a report to us, but was concerned about their results because the students got confused while doing the testing. However, the teachers were very pleased that students had a chance to try the test kits and work on an authentic research project.

The authors were little disappointed with the results at Suksasongkror Bangkruay School. This school potentially was the most interesting, since many of the students are from hill tribes in northern Thailand, near the headwaters of the Chao Phraya River. They had two challenges at this school. First, because the students generally come from a remote area in Thailand, their primary school training did not prepare them well for this type of study. More importantly, the teacher assigned to this project was not as dedicated as we had hoped in guiding the students.

(a)

รายงานสรุปผลโครงการศูนย์เรียนรู้แม่น้ำเจ้าพระยา

หลักการและเหตุผล

โรงเรียนคณะราษฎร์บำรุงปทุมธานี เป็นโรงเรียนที่ตั้งอยู่ริมฝั่งแม่น้ำเจ้าพระยาซึ่งเป็นแม่น้ำสายสำคัญของประเทศ เป็นแหล่งอารยธรรมและวัฒนธรรมของชาติ ก่อเกิดประเพณีต่างๆ มากมายเกี่ยวกับสายน้ำ แม่น้ำเจ้าพระยาได้หล่อเลี้ยงผู้คน อำนวยประโยชน์ต่อคนในชาติอย่างอเนกอนันต์ ทั้งใช้ในการอุปโภคบริโภค การคมนาคม แหล่งอาหาร และการท่องเที่ยว แม่น้ำเจ้าพระยาจึงเป็นแหล่งเรียนรู้ที่สำคัญของเยาวชน ที่ทำให้นักเรียนมีจิตสำนึกในการอนุรักษ์และพัฒนาสิ่งแวดล้อมที่เกี่ยวพันกับแม่น้ำเจ้าพระยา และยังมีจิตอาสาในการช่วยกันอนุรักษ์แม่น้ำเจ้าพระยาให้มีสภาพแวดล้อมที่ดีสืบไป

ดังนั้นทางโครงการเรือไตรทศ ร่วมกับมหาวิทยาลัยนิวยอร์คบัฟฟาโล่ เสตท คอลเลจ นำโดยศาสตราจารย์ เท็ด ทัคเคิ่ล และศาสตราจารย์ คิม เออไวน์ ให้การสนับสนุนอุปกรณ์ในการตรวจสอบคุณภาพน้ำในแม่น้ำเจ้าพระยาให้แก่โรงเรียนคณะราษฎร์บำรุงปทุมธานี และจัดกิจกรรมการเรียนรู้ที่เกี่ยวข้องกับแม่น้ำเจ้าพระยาต่างๆ ในระยะเวลาหนึ่งปีการศึกษา ทางโรงเรียนได้ส่งผลการตรวจสอบคุณภาพในแม่น้ำเจ้าพระยาเดือนละครั้ง ตั้งแต่เดือนมีนาคม 2555-เดือนมีนาคม 2556 ให้แก่โครงการเรือไตรทศทุกๆเดือน

วัตถุประสงค์

1. เพื่อให้นักเรียนมีแหล่งเรียนรู้ที่เกี่ยวข้องกับแม่น้ำเจ้าพระยา
2. เพื่อให้นักเรียนมีจิตสำนึกในการอนุรักษ์และช่วยกันรักษาแม่น้ำเจ้าพระยา

Fig. 6.10 Reports from Kanarajbumroong Patumthani School (**a**) and Sriboonyanon School (**b**)

(b)
โครงการผู้พิทักษ์สายน้ำในโครงการเรือไตรทศร่วมกับ Buffalo State College New York University U.S.A.

มอบอุปกรณ์วิเคราะห์คุณภาพน้ำและอบรมนักเรียนแกนนำ

นักเรียนห้องเรียนพิเศษชั้น ม. 4 โรงเรียนศรีบุณยานนท์ ปีการศึกษา 2554

Fig. 6.10 (continued)

An important takeaway from the project is the necessity of developing a strong collaborative interaction that provides training and guidance for the teachers and students, but also not surprisingly, a committed teacher cohort to implement the program.

While the chapter has provided an extensive description of the thinking behind the program, the design considerations, the actual learning experience, and the links to geographical and environmental education, the program can benefit from a more structured evaluation of students' beliefs and attitudes in future. For now, the chapter presents key challenges and learning points that will provide an exemplar of how field-based learning in geography and environmental education can be conducted.

The journey to Education for Sustainable Development will not be an easy one. Challenges to implementing the program also included the record flood of 2011 that closed schools and delayed the start of sampling, as well as the theft of analytical kits. The schools involved in the project are not elite schools, yet the majority of teachers viewed the project as positively impacting student learning and critical thinking skills. It is precisely these types of schools where intervention through authentic, non-formal learning can have the biggest impact on education in Thailand.

Acknowledgements The authors would like to express our sincere gratitude and deep appreciation to Mr. Ted Turkle for his vision in initiating and supporting this program through the SUNY Research Foundation. Special thanks to Mr. Kim Irvine to inspire me to work on this project. Furthermore, we would like to thank teachers and students from all the project schools who always worked hard to get the test result and for their warm coordination. I also would love to thank Lynda Rolph who supported me throughout this project. Finally, I also would like to thank all those who have been a part of this research who have not been mentioned.

References

American Public Health Association (APHA). (2005). *Standard methods for the examination of water and wastewater* (21th ed.). Washington, DC, USA: American Public Health Association (APHA).
Bhargava, D. S. (1983). Use of a water quality index for river classification and zoning of Ganga River. *Environmental Pollution (Series B), 6,* 51–67.
Bordalo, A. A., Nilsumranchit, W., & Chalermwat, K. (2001). Water quality and uses of the Bangpakong River (Eastern Thailand). *Water Research, 35*(15), 3635–3642.
Boyle, A., Maguire, S., Martin, A., Milsom, C., Nash, R., Rawlinson, S., et al. (2007). Fieldwork is good: The student perception and the affective domain. *Journal of Geography in Higher Education, 31*(2), 299–317.
Brown, R. M., McClelland, N. I., Deininger, R. A., & Tozer, R. G. (1970). A water quality index —do we dare? *Water and Sewage Works, 117,* 339–343.
Brundiers, K., Wiek, A., & Redman, C. L. (2010). Real-world learning opportunities in sustainability: from classroom into the real world. *International Journal of Sustainability in Higher Education, 11*(4), 308–324.
Cude, C. G. (2001). Oregon water quality index: A tool for evaluating water quality management effectiveness. *Journal of the American Water Resources Association, 37*(1), 125–137.
de Haan, G. (2006). The BLK '21' programme in Germany: a 'Gestaltungskompetenz'-based model for Education for Sustainable Development. *Environmental Education Research, 12*(1), 19–32.
Dojlido, J., Raniszewski, J., & Woyciechowska, J. (1994). Water quality index—application for rivers in Vistula River Basin in Poland. *Water Science and Technology, 30*(10), 57–64.
Dummer, T. J., Cook, I. G., Parker, S. L., Barrett, G. A., & Hull, A. P. (2008). Promoting and assessing 'deep learning'in geography fieldwork: An evaluation of reflective field diaries. *Journal of Geography in Higher Education, 32*(3), 459–479.

Food and Agriculture Organisation of the United Nations (FAO). (2016). *AQUASTAT main country database*. Retrieved February 24, 2016, from AQUASTAT: http://www.fao.org/nr/water/aquastat/dbase/index.stm.

Gilbert, T. W., Behymer, T. D., & Castañeda, H. B. (1982). Determination of dissolved oxygen in natural and wastewaters. *American Laboratory, 14*(3), 119–134.

Haigh, M. J. (2006). Promoting environmental education for sustainable development: The value of links between higher education and Non-Governmental Organizations (NGOs). *Journal of Geography in Higher Education, 30*(2), 327–349.

House, M. A., & Ellis, J. B. (1987). The development of water quality indices for operational management. *Water Science and Technology, 19*(9), 145–154. Retrieved November 19, 2012, from http://www.watersfoundation.org.

Irvine, K. N., Rossi, M. C., Vermette, S., Bakert, J., & Kleinfelder, K. (2011). Illicit discharge connection and elimination: Low cost options for source identification and trackdown in stormwater systems. *Urban Water Journal, 8*(6), 379–395.

Jickling, B. (1994). Why I Don't Want My Children to Be Educated for Sustainable Development. *Trumpeter, 11*(3), 114–116.

Kern, E. L., & Carpenter, J. R. (1984). Enhancement of student values, interests and attitudes in Earth Science through a field-oriented approach. *Journal of Geological Education, 32*(5), 299–305.

Komori, D., Nakamura, S., Kiguchi, M., Nishijima, A., Yamazaki, D., Suzuki, S., ... & Oki, T. (2012). Characteristics of the 2011 Chao Phraya River flood in Central Thailand. *Hydrological Research Letters, 6*, 41–46. Published online in J-STAGE http://www.jstage.jst.go.jp/browse/HRL, https://doi.org/10.3178/hrl.6.41.

Lok, J. J. (2014). *Comparison of water quality test kits (LaMotte, CHEMetrics and Test Strips) at Bishan Park and Kallang River Canal* (41 p). AAG401 Undergraduate Final Year Project Report, National Institute of Education, Singapore.

McGuinness, M., & Simm, D. (2005). Going global? Long-haul fieldwork in undergraduate geography. *Journal of Geography in Higher Education, 29*(2), 241–253.

Mitchell, M. K., & Stapp, W. B. (1995). *Field manual for water quality monitoring an environmental education program for schools* (9th ed.). Ann Arbor, MI: Green Project.

Mogensen, F., & Schnack, K. (2010). The action competence approach and the 'new' discourses of education for sustainable development, competence and quality criteria. *Environmental Education Research, 16*(1), 59–74.

National Aeronautics and Space Administration (NASA). (2016). *Nakhon Sawan Gauging Station data summary*. Retrieved February 24, 2016, from Earth Data: http://daac.ornl.gov/rivdis/STATIONS/TEXT/THAILAND/888/SUMMARY.HTML.

Ng, Y. W. A. (2014). *Evaluation and comparison of water quality test kits for the purposeful use in the lower secondary geography syllabus in Singapore* (75 p). AAG401 Undergraduate Final Year Project Report, National Institute of Education, Singapore.

Ongsakul, R., & Sajor, E. E. (2006, November). Water governance in mixed land use: a case study of Rangsit Field, peri-urban Bangkok. In proceedings of the Regional Conference on Urban Water and Sanitation in Southeast Asian Cities, AIT (pp. 329–340).

Patarasiriwong, V. (2000). Water quality of the Rangsit Prayoonsak Canal. *Kasetsart Journal (Social Sciences), 21*, 109–117.

Pesce, S. F., & Wunderlin, D. A. (2000). Use of water quality indices to verify the impact of Cordoba City (Argentina) on Suquia River. *Water Research, 34*(11), 2915–2926.

Price, J., Chaosakul, T., Surinkul, N., Bowles, J., Rattanakul, S., Pradhan, N., et al. (2011, December). Surface water quality and risk analysis in a peri-urban area, Thailand. In proceedings of the 9th International Symposium on Southeast Asian Water Environment, Bangkok, Thailand (pp. 399–406).

Prüss-Ustün, A., Bartram, J., Clasen, T., Colford, J. M., Cumming, O., Curtis, V., ... & Freeman, M. C. (2014). Burden of disease from inadequate water, sanitation and hygiene in low-and middle-income settings: a retrospective analysis of data from 145 countries. *Tropical Medicine & International Health, 19*(8), 894–905.

Reid, A. (1990). An "age of commerce" in Southeast Asian history. *Modern Asian Studies, 24*(1), 1–30.

Simachaya, W. (2003, November). A decade of water quality monitoring in Thailand's four major rivers: The results and implications for management. In Proceedings of the 6th International Conference on the Environmental Management of Enclosed Coastal Seas, Bangkok, Thailand.

Sitthithanyakij, P. (2012). Chao Praya River Legend, Thailand. Siam Knowledge.

Smith, G. (1999). Changing fieldwork objectives and constraints in secondary schools in England. *International Research in Geographical and Environmental Education, 8*(2), 181–189.

Steele R. (2011). *Becoming a compass school, train the trainer handbook.*

Suwanarit, A. (2012). Building living landscapes—Future developments must respect nature. *Asia-Pacific Housing Journal, 6*(18), 17–24.

Tilbury, D. (2011). Higher education for sustainability: A global overview of commitment and progress. *Higher Education in the World, 4,* 18–28.

Tilbury, D., & Wortman, D. (2008). How is community education contributing to sustainability practice? *Applied Environmental Education and Communication, 7(3),* 83–93.

Tiptus, P. (2000). Settlement in the Central Region of Thailand. In *Proceedings of the International Conference: The Chao Phraya Delta: Historical Development, Dynamics and Challenges of Thailand's Rice Bowl, Thailand.*

UNESCO. (2011). *Education for sustainable development.* UN Decade of Education for Sustainable Development 2005–2014. UNESCO

UNESCO. (2006). *Chao Phraya River Basin, Thailand.* Retrieved February 24, 2016, from World Water Assessment Programme: http://webworld.unesco.org/water/wwap/case_studies/chao_phraya/chao_phraya.pdf.

United Nations University. (2016). *Integrated water resources management.* Retrieved February 24, 2016, from UN Water Learning Centre: (http://ocw.unu.edu/international-network-on-water-environment-and-health/introduction-to-iwrm/introduction-to-iwrm.zip/view.

Vare, P., & Scott, W. (2007). Learning for a change exploring the relationship between education and sustainable development. *Journal of Education for Sustainable Development, 1*(2), 191–198.

Venkataraman, B. (2009). Education for sustainable development. *Science and Policy for Sustainable Development, 51*(2), 8–10.

Villiers, J. (1999). Portuguese and Spanish sources for the history of Ayutthaya in the sixteenth century. *Journal of the Siam Society, 86*(Parts 1 and 2), 119–130.

Wals, A. E. J., & Kieft, G. (2010). Education for sustainable development research overview. *SIDA Review* 2010:13. Sweden.

Waters Foundation. (2016). *Systems thinking in schools.* Retrieved February 24, 2016, from Waters Foundation: http://watersfoundation.org/.

Wills, M., & Irvine, K. N. (1996). Application of the National Sanitation Foundation water quality index to the Cazenovia Creek pilot watershed management study. *Middle States Geographer, 29,* 95–104.

น. ณ ปากน้ำ (1989). แบบแผนบ้านเรือนในสยาม. (10–13). Thailand. Aksornsumpun Publishing.

World Commission on Environment and Development (1987). *From One Earth to One World: An Overview.* Oxford: Oxford University Press.

Supitcha Kiatprajak is a Member of the Three Generation Barge Program team and has worked with communities and wildlife in Thailand. She has obtained BA in Humanities and English from the University of the Thai Chamber of Commerce and Master of Education in Environmental Education from Mahidol University. She would like to always share her experiences and help everyone to learn to sustain our natural resources, other creatures, and our world. She loves art, music, traveling, and all kinds of outdoor adventures.

Lynda graduated with a BEd and taught for nine years in London, before moving to Thailand to teach primary science in an English-medium Thai school. She has held different areas of responsibility during her many years with Traidhos Three Generation Community for Learning, Thailand. She worked with the Girl Guide Association of Thailand to set up preschools in northern villages and has led groups in camping and outdoor activities.

Chapter 7
Paradigm Shift in Humanities Learning Journey

Marc Teng and Kah Chee Chan

Abstract Teachers are increasingly being tasked to conduct learning journeys for schools. A term specific to the Singapore context, learning journeys refers to field-based learning, at a specific site or sites. The concept of journey is used as the entire learning experience can be designed around a trip or a journey. This is especially true for a humanities subject such as geography that requires the context of learning in the lived environment, and sometimes in an overseas context. On one hand, students might reject the programs if they are too instructional based and academic. On the other hand, learning points would be missing if there is an over-emphasis on activities. Thus, the salience in every learning journey lies in defining the learning objectives and outcomes of the students. While field-based learning can be conducted within the context of a disciplinary subject, such as geography, there are some broad considerations for organizing learning in the field. Teachers could initiate the program with an end in mind as better outcomes could only be achieved if the whole idea of the trip is latched onto this concept. It is imperative that teachers revisit and determine the learning outcomes and learning objectives before the implementation of a learning journey. Upon delineating the aims of the learning journey, teachers would have to choose among the different tools in the learning journey, in this case, the different destinations or activities to elicit the intended outcome that is relevant and engaging for the students. For monitoring purposes, the whole process can be re-examined using a Balanced Scorecard to ensure that the intended outcomes are achieved. The Balanced Scorecard will also serve as a feedback loop to refine subsequent learning journey processes which will in turn provide efficient directions for future field trips.

M. Teng (✉)
STA Travel, Singapore, Singapore
e-mail: Marc.Teng@statravel.com

K. C. Chan (✉)
Wholistic Learning Consortium, Singapore, Singapore
e-mail: wholistic_kcchan@singnet.com.sg

© Springer Nature Singapore Pte Ltd. 2018
C.-H. Chang et al. (eds.), *Learning Geography Beyond the Traditional Classroom*,
https://doi.org/10.1007/978-981-10-8705-9_7

Introduction

Overseas education travel or learning journeys are becoming more predominant in government schools due to the government initiatives to encourage more overseas travel for MOE schools under the Global Outreach program. The learning journey Programs was initiated by MOE in 1998 with the aim of promoting overseas travel to the students in Singapore. The Ministry of Education and the different education institutions in Singapore have the vision in providing the students with greater exposure on foreign cultures, traditions, and different aspects of the foreign society through such programs. Many schools in Singapore have responded to such calls for actions accordingly and the numbers of school trips overseas have increased since 1998 due to such patronage by the Ministry of Education. The government has always been supportive with such the learning journey Programs and they have even increased the funding for individual schools so that each and every child could be sent overseas for learning journeys. This could be seen through the establishment of the "International Fund" which was set up in 2008 to provide the schools in Singapore with $22 million for their internationalization efforts (Durai and Mao 2010). Even though the learning journeys are not entirely funded by the schools or Ministry, they have provided a substantial amount of subsidy to assist the students in Singapore with these programs and this has subsequently led to the success of this program.

The learning journeys are initiated in 1998 as 'part of a process of active learning to expose pupils to the sights and sounds of different environments outside the classroom. It is also an important learning experience, essential, and integral to the educational process' (Ministry of Education, Singapore 1998). Learning journeys are also undertaken by education institutions in other countries but the chapter will focus on the examples from Singapore, with a view to provide discussion points for learning journeys conducted by other countries.

As most education institutions usually lack the necessary logistical support and contacts to enable them to plan and execute such trips on their own, there is a need to rely on vendors in the travel industry to assist them in these overseas trips. While the responses to most of these trips conducted by these travel vendors are mostly positive, learning journeys by travel vendors are generally viewed by teachers and students no differently from leisure trips as it does not always delineate and clarifies the objectives and outcomes of such trips aptly.

As a point of reference, educators would be able to synchronize the learning journeys more effectively through the use of a Strategize-Implement-Outcome (SIO) model (see Fig. 7.1). The SIO model would provide a focal point for educators and other professionals, as it would function as a tool to examine and scrutinize the effectiveness of a learning journey. Educators would identify the learning objectives and outcomes hand in hand so as to achieve the desired results. Upon establishing these two areas of concern, educators could then select a series of deliverables so as to bridge the current reality to their perceived outcome. Consequently, the verification of the program could be reinforced through the use of a learning journey Balanced Scorecard to measure the success of the trip itself.

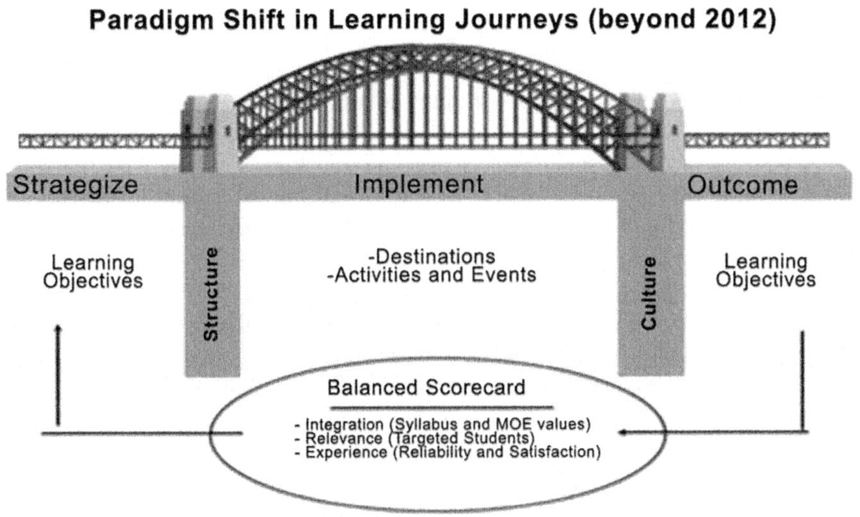

Paradigm Shift in Learning Journeys (beyond 2012)

Fig. 7.1 SIO model for learning journeys

Unpacking the SIO Model

See Fig. 7.1.

Strategic Implications: Learning Objectives

The focus in every learning journey lies in defining the learning objectives and outcomes of the students. Teachers could initiate the program with an end in mind as better outcomes could only be achieved if the whole idea of the trip is latched onto this concept. This theory is widely postulated by academics such as Stephen Covey in his eminent work the "Seven Habits of Highly Effective People" which espouses the concept of having the end in mind (Covey 2004). Teachers who are planning learning journeys could similarly adopt a similar strategy by conceptualizing the end result prior to the commencement of the program. Hence, whether the objectives are to inculcate values in the student, impart a series of skill sets or to internalize a theme from the syllabus, the learning objectives for each trip should be clearly defined. It is important to take reference to the learning objectives that are required of the student within the context of the school curriculum. In the topic of Tourism, for instance, the national syllabus in Singapore requires students to be able to explain how tourism can be made sustainable, and compare the roles of various groups in taking care of tourist areas. With these objectives in mind, the selection of the sites and the design of the activities and resources to support these activities will become relevant to the students' learning.

Implementations: Destinations

The tools for an educator during a learning journey would naturally be the environment and artifacts that are on display as such experience are something that cannot be replicated in a classroom. Hence, the selection of the right destinations would be imperative for the educators as it would define the learning outcomes of the students. Selecting the right environment is important as the field is an important place to gather, analyze, and dispense data for tactical or strategic purpose (Choy 2011). To begin, educators may want to begin with a macro framework linking the core competencies of the various countries to the different themes or values in the syllabus or curriculum. Thus, an educator would want to initiate a learning journey to Berlin if the aim or outcomes for the students is to learn about the rise of Hitler or the Cold War. The educator could then subsequently narrow this down to the few destinations within Berlin so as to maximize learning within the students. Alternatively, Taiwan might be an excellent learning location for the students to study more about coastal erosional features or rock formation. The educator could select Wanli, Yehliu, for this purpose. The whole process would be further reinforced through the designed activities and relevant academic materials jointly developed by the educators and vendors.

Expected Outcome of the Students

This component would have been taken into consideration by the educators as it would have been covered during the initialization stage. By having a clearer and more tangible form of assessment during and after the program, educators would be able to scrutinize and keep track of the learning outcomes more accurately. Relevant supplementary materials during the learning journey which assess the students' understanding of a particular topic at a particular destination would hence be a necessity to determine whether the learning outcome has been attained.

Learning journey Balanced Scorecard

In adopting the SIO model, it is also important to bear in mind the need to balance between making the activities on a learning journey fun and ensuring the academic rigor of the program by engaging the students in meaning instruction and learning. The *Balanced Scorecard* is a strategic performance tool developed by Robert Kaplan for the managers of different organizations to assess the viability of their programs (Kaplan and Norton 1996). While the original Balanced Scorecard consists of the Learning/Development, Internal Processes, Customer Satisfaction, and Financial aspect, the learning journey Balanced Scorecard in this instance would measure the workability and the viability of a school learning journey.

The learning journey Balanced Scorecard would consist of three components instead of the originally four sections. The section on Learning and Development would be modified to become the Integration of Syllabus/Curriculum section, which will form the base of the whole learning journey. Likewise, the internal processes would be represented by the relevance of the learning journey which would explore and scrutinize the processes of the whole learning journey. The third part which characterizes customer satisfaction in the original learning journey rubrics would be replaced by the section on Users experience in this category. The users here would generally refer to the end consumers, whom consist of the participating teachers and the students in general. Using the example of a learning journey to be organized for the topic of tourism in Hanoi, the Integration of Syllabus/Curriculum section will need to address what the key aims of the topic on Tourism are in the current curriculum. The relevance of the learning journey needs to refer to the specific learning outcomes for students that are aligned to the curriculum and the student and teachers learning experience in the field should be taken into consideration, such as through the activities planned on the learning journey. This could include understanding the language differences and the logistical requirements of the site to make sure that the field study can be carried out smoothly.

Integration of the Curriculum

As stated earlier, learning journeys that do not conform to the curriculum or the greater MOE framework is at best a leisure trip. Hence, it is a necessity for vendors and educators to collaborate on the destinations and the learning outcome so as to provide a more pertinent trip for the students. While it may be cumbersome for educators and vendors to collaborate for every of such trips, the ministry could step in by roping in the various educational institutions and travel vendors for focus groups sessions to strategize, synchronize, and develop a pool of resources for the schools so that educators are able to learn more about the achievements and pitfalls of different learning journeys and plan subsequent field trips more efficiently.

Relevance of the Learning Journey

Mapping the learning objectives and outcome to the itineraries of the learning journeys would only signify the intent to make the learning journey more relevant for the schools and students. The process on how the entire trip is carried out is thus worth highlighting and examine in the learning journey Balanced Scorecard as it should serve as a gauge on the significance of these trips. Thus measures such as the ability for the local guides to connect with the students and the ability to match the materials to the learning objectives should constitute as one of the components in the Scorecard as well. Vendors like educators should also employ a myriad of activities to teach a class with mixed abilities (Toh 2005). In addition to this, as aptly pointed out by Cheong, multiple stakeholders should be engaged and diverse resources should be brought, so as to facilitate students' multiple and sustainable development (Cheong 2012). Indeed, the internal efficacy of both local and foreign travel vendors will drive the significance and overall experience of the students. Such efficiency could only be derived through the professionalism of the vendors which is acquired through due diligence by travel vendors during the recruitment process and consistent training for its staff.

User Experience for the Learning Journey

The user (teachers and students) experience remains largely important to the vendors and the school because these experiences from these trips would serve as subsequent signposts or markers for the schools to continue with a particular travel vendor for future learning journeys. Assessment of such experience is relatively easier as compared to the other two components as participants are more willing to share about the conditions of the accommodation, food, and their travel experience more readily through feedback sessions. The MOE post-learning journey review or feedback is one of the many avenues for such feedback as well.

Discussion and Conclusion

While the SIO model provides a conceptual framework while designing learning journeys, there needs to be several key enablers before learning journeys can be successfully conducted. These enablers will include structural or institutionalized support and the culture of the school, among other things. The institutionalized support by the education ministry and government is without a doubt one of the major pillars which supports the learning journey initiatives of any MOE school. The absence of the funding would render such efforts financially unmanageable as schools and organizations would have to seek funding from external parties. Structurally, the ministry is also required to set a general direction and directives for the schools so that these trips could encompass the larger MOE framework and the strategic thrusts that schools have to align to.

Apart from the patronage of the ministry, the eventual success of a learning journey would require a bottom-up approach as the program would require commitment from the participants, namely the educators and the students also. On one hand, the vendors would need to ensure that such trips for the students are engaging and relevant for their consumers so as to sustain such a program; on the other hand, educators would need to pique up the interests through pre-trips activities to ensure the sustenance of the motivation from the students prior to the learning journeys.

By and large, learning journeys in Singapore are still in its infancy as these initiatives were only introduced less than 15 years ago. Currently, the line between leisure travel and learning journeys is still indistinct and greater collaboration could be initiated between education institutions and travel vendors to forge a superordinate objective, which would benefit the students in the long run. Teachers would be at a disadvantage if they take on the process of education in silo as they need to build collaborative relations with people in the industry and community who can provide them and their students with learning opportunities in the changing environment outside the classroom (Khong 2005). The implementation of the SIO model does have its limitation, as it would require much effort from the educators at the initial stage. Apart from this, the adoption of the SIO model might also encounter resistance initially due to a change in corporate culture. Henceforth, it is up to the school management and department heads to communicate and empower the individuals about the benefits before the implementation so as to maximize learning among the students. Nevertheless, teachers need to be able to make evidence-informed decisions about their practice, and in this case, the SIO model supports teachers in making decisions during the planning and implementation of field study.

References

Cheong, C. (2012). Teachers for new learning: reform and paradigm shift for the future. In O. Tan (Ed.), *Teacher education frontiers: International perspectives on policy and practice for building new teacher competencies* (p. 97). Cenage Learning.

Choy, W. (2011). Globalization and the dynamic educational environment in Singapore. In W. Choy, & C. Tan (Eds.), *Education reform in Singapore: Critical perspectives* (pp. 6–7). Prentice Hall, Pearson Education South Asia.

Covey, S. (2004). *Seven habits of highly effective people*. Free Press.

Durai, J., & Mao, Z. (2010, April 30). *School trips offer lessons abroad*. Retrieved September 14, 2012, from AsiaOne: http://www.asiaone.com/print/News/Education/Story/A1Story20100428-212943.html.

Kaplan, R., & Norton, D. (1996, January). *Using the balanced scorecard as a strategic management system*. Retrieved September 7, 2011, from Harvard Business Review: http://www.hbr.org.

Khong, L. (2005). School-stakeholder partnerships: Building links for better learning. In P. Ng, & J. Tan (Eds.), *Shaping Singapore's future* (p. 117). Singapore: Prentice Hall, Pearson Education South Asia.

Ministry of Education, Singapore. (1998, February 24). Launch of learning journeys. Retrieved September 12, 2012, from Ministry of Education: http://www.moe.gov.sg/media/press/1998/980224.htm.

Toh, C. (2005, March). Trigger the Senses. *Ideas on teaching* (p. 34). Centre for development of teaching and learning. Retrieved from http://www.cdtl.nus.edu.sg/Ideas/iot86.htm

Marc Teng June Lean As a former humanities educator, Teng June Lean, Marc is the Education Groups Manager of STA Travels, Singapore. He is also the international liaison for various conferences like Higher Education Forum (HEF), International Research Symposium on Engineering and Technology (IRSET), and International Symposium on Education and Social Science (ISESS).

Kah Chee Chan is the honorary treasurer for the Wholistic Learning Consortium. His doctoral research work in holistic management recommends the adoption of a holistic business and marketing approach by managing five interrelated relationships for sustainable competitive success.

Part III
Teaching and Learning Geography with Information and Communication Technology

The advancement of new technology has brought about significant changes in our lives. The potential of incorporating this technology in teaching and learning to improve the delivery of teaching and facilitate learning beyond the traditional classroom to engage and enrich students is especially critical in geography. This part includes examples of the type of technology incorporated into experiential and inquiry-based learning in Singapore and how technology can potentially improve the learning experience. This is particularly useful in field inquiry where the work of students may be shared using technology for analysis and therefore learning will not be limited by the traditional constraints of space, time and place.

Chapter 8 provides a comprehensive use of a proprietary mobile phone app, NIE mGeo, where students can observe field feature, track it on the mGeo application and field sketch on the same device. This chapter provides a background to the development of this mobile application. Information shared on the app can be analysed by different groups at different locations to harness the potential in mobile technology to manage a total learning environment. This will help educators who are trying to manage the curriculum and holistic learning experience in the limited curriculum time where the geo-referenced data can be used for analysis.

Part III continues with Chap. 9 that aims to add value to the discourse on incorporating technology into practice with a constructivist approach where students define meaning in the information they encounter and learn in the process. The chapter highlights the importance of the role of the educator in the planning of the use of technology in geography inquiry learning and the support given to students for the construction of knowledge. The chapter discusses in detail the usage and representation of data on mobile devices, collaborative inquiry processes and the facilitator's role and pedagogical responsibility in the collaborative processes. It also recognises the exciting opportunities ahead for teaching geography where key geographical concepts such as space, place and scale can be engaged in the classroom through the use of technology.

Part III concludes with the case study of using high-speed mobile technology in the classroom to facilitate geography lessons where technology allows students to learn about the world beyond physical space restrictions. In addition to the opportunities, the chapter also discusses about the challenges of incorporating technology in practice, such as time, effect and resource constraints that might affect the decisions of employment of high-speed mobile technologies. However, the benefits that students reap from these classes outweigh the costs where the students now learn better in authentic learning settings where they have to solve real-world problems (Barron and Darling-Hammond 2008).

The three chapters provide the pedagogical considerations of the use of mobile technologies in geographical fieldwork in terms of the conceptual thinking behind developing the mobile application, its use for collaborative learning in the field and the consideration of its affordance to take the students beyond the constraints of physical space. While there are challenges in using mobile applications for learning geography, the potential of the opportunities to improve student learning far outweighs the costs, and in this rapidly changing world, it is critical that students develop the holistic geographical skills to solve real-world problems.

Reference

Barron, B., & Darling-Hammond, L. (2008). *Teaching for meaningful learning: A review of research on inquiry-based and cooperative learning.* Book Excerpt. George Lucas Educational Foundation.

Chapter 8
Authentic Learning: Making Sense of the Real Environment Using Mobile Technology Tool

Kalyani Chatterjea

Abstract Formal learning, though usually conducted within the four walls of classrooms, with or without technology, is not complete, unless connected to the authentic environment. This has made ground truthing essential. With present-day technology, we can measure an entire catchment and understand its capacities, impact on a region and mankind's positive or negative influences. But without verification by ground truthing, our data from space are hypothetical at best. This is where fieldwork becomes unquestionably an essential part of learning. However, learning through fieldwork is not just about having a direct experience of the locational environment, but also about making educated judgement about the location, based on observations and measurements of desired parameters, to make sense of the environment, to be locationally aware, contextually rich, and to be able to relate the two in a way that unravels the uniqueness of the subject. Today's many technological tools can provide the required support for doing such a task, but it requires a modest investment of time and resources to produce something academically substantial. Mobile technology, however, has managed to bring the many previously impossible tasks together and made them not just possible but also pervasive and affordable, not just to the elite few but to the general learner groups. For this reason, in the present learning arena, using the mobile technology is not just riding the technological bandwagon, but an essential vehicle to reach out to the learners far and wide, to empower the masses, to encourage even the 'not-so-initiated' learners to think, to excite the 'already-initiated' learners to look further afield and generally to expand the horizons of learning. Field-based authentic learning, though desired, is often hard to execute due to big student groups, lack of curriculum time and the inevitable need to gather huge database for any in-depth research to achieve academically sound outcomes. Because of the operational constraints, many a time field-based learning is done to fill in some gaps in the learning strategies but cannot be incorporated in a comprehensive learning outcome. The chapter focuses on authentic learning through fieldwork and also how applications on mobile phones are used to not just help in field observations but also

---

K. Chatterjea (✉)
National Institute of Education, Nanyang Technological University, Singapore, Singapore
e-mail: kalyani.c@nie.edu.sg

how mobile technology is used in managing the total learning environment, starting from field data collection to post-field data organization, and analysis, thus completing the whole learning circle to create a knowledge base on the intended topic. It will focus on how even large groups of students can be engaged in doing authentic field-based study and also do in-depth analysis through collective database development and socially negotiated knowledge. The chapter also looks at how mobile technology can be used to conduct large group learning in authentic environments and how learning is helped by using modes with least learning curves and without additional capability requirement. There are examples from actual work done by students in remote areas, using mobile technology, but with little other infrastructural support, to collectively develop academically sound data sets and then analysing these to achieve target learning outcomes, without impinging excessively on the given curriculum timeframe.

Introduction

Authentic learning is touted as the most effective way of conveying the key aspects from relevant experiences to one's learning environment and the most efficient way of transforming everyday occurrences to enhance our learning experiences. By being spatially situated and being connected with life, authentic learning makes us aware of the processes around us. It helps us to make sense of the processes around us and also enables us to relate classroom teaching to the various processes around us, thus providing the much-needed relevance to what we learn in the classrooms. Having the source in the places of occurrence, authentic learning environments help us create new knowledge with the help of the processes around us, by integrating classroom-based conceptual knowledge with the observed processes, supporting a hypothesis or challenging it, to create a new set of understanding and a higher level of cognition. Therefore, we can say that authentic learning completes the process of knowledge acquisition, by providing the learner the opportunity to not just practically experience the theoretical knowledge, but also providing the opportunity to understand first-hand, validate, alter and finally assimilate to form a personal learning experience, which is totally designed by the learner. So through authentic learning, we can go from a teacher-presented knowledge set to one that is entirely learner-centred knowledge creation. While we aim to provide the learners a greater involvement in their learning, authentic learning, therefore, can and does play a monumental role in making learning personal and therefore relevant. Education or learning, therefore, cannot be or should not be without some authentic learning experiences.

Since authentic learning provides relevance to the real world, it goes without saying that it has to involve some direct experiences from the real world. This makes field experiences play a central role. Fieldwork for getting information directly from the source, ground truthing for spatial information, primary data collection from relevant sources are some ways of getting real-world scenarios into the learning arena, and all of these would involve some kind of field experiences. In today's teaching and learning scenarios, therefore, increasing emphasis is placed on getting in touch with the real world—be it is schools or higher education institutions, more and more project-based courses are being introduced, more institutional attachments are emphasized, more direct experiences are being encouraged as a part of this authentic learning experience. In disciplines where the world at large is the learning arena, more and more students are being taken out of classrooms, to gain knowledge directly from the world outside—this is the value of fieldwork that cannot be over-emphasized.

Integrating direct field experience in the regular curriculum requires a fair amount of work on the part of the teacher and the course coordinator. From a learning point of view, the learner has to fulfil certain objectives. The fieldwork involves (a) observation; (b) data collection through use of various instruments; (c) data recording; (c) searching for patterns, requiring a sizeable data set; (d) scope for personal exposition on the observation; (e) security of observed and recorded data sets. Collection of data from an authentic environment, by itself, does not result in any knowledge creation, unless the collected data are further analysed. Thus, the process of learning from an authentic environment continues into post-field session, when real-life experiences are analysed and synthesized to create a new knowledge base. In a traditional setting, such processes are conducted and followed in many places only through a lot of concerted zeal and many pitfalls, as there are too many components that can and do go wrong and because the traditional classroom focus on education does not normally allow adequate curriculum time for such a time-consuming exercise. An authentic learning environment, though much aspired, falls short of being just a few instances of sporadic effort in providing effectiveness in the learning experience. Zealous systems still labour through it, and increasingly, we observe field exposures as part of holiday/term break activities in schools. These vacation-time activities, however, either take precious time away from learners who need some time to enjoy a real break or do not truly become academically engaging enough to deliver the real learning point, as expected. In order to integrate authentic field experiences within the curriculum time, a system needs to be integrated enough and easy enough to follow and pursue, without being too demanding. One way of doing this is to use technologies that are already pervasive among young learners, to overcome some hurdles as well as to enhance the way we learn. Mobile technology is one such technology, which is not just extremely pervasive in today's society but also one that is fairly affordable and definitely without a learning curve. A mobile device, being small, also scores a point in being a device that is handy in field situations. With these features and many others, a mobile device has now become one affordance that the young learners automatically take to and integrate in their stride in life. This makes it an effective and useful device to latch on to, to use in learning.

While there have been a few chapters in the book that has already described how NIE mGeo has been used in the field, this chapter will discuss its development and how it incorporates the features of authentic learning from field-based situations and integrates conceptual knowledge to real-world processes, to enhance the way learners can integrate knowledge and in the process create personalized knowledge base. The application has been used by more than 2700 students in many schools across Singapore and Indonesia and also in university and among teachers in Singapore, to create new customized knowledge, relevant to the respective curriculum. The chapter looks at the tool developed to be used on mobile phones that has built-in features to facilitate intensive fieldwork and substantial collaborative data set development, without making such a task too difficult and time-consuming. The paper also examines how, through the use of the developed mobile application, fieldwork can contribute towards collaborative knowledge building and individual knowledge creation and meaning-making through analysis of collaborative data related to some authentic environments.

Objectives and Scope of Research

To be useful and purposeful, an authentic learning strategy has to be data oriented, quantitative or qualitative, to ensure availability of empirical evidence to support the learning. Increasingly, there is a trend to include spatial relevance to information, to make it more authentic and to provide both temporal and spatial reference to information. Such field-derived data also need to be in substantial quantity to be useful for analysis. In other words, an effective field research should have features such as photographic evidences, geo-spatial reference, textual as well as numerical data, and a large number of data sets. To put all these together in a field-based research requires many platforms, such as various data generating instruments, series of data recording sheets, camera, some kind of collaboration to ensure that enough primary data are gathered from the field. Handling of multiple instruments, formats and devices can be cumbersome and may hinder learners from achieving a seamless learning experience. Such operational limitations led to the conceptualization of a mobile application that is designed to provide support for field-based authentic learning and being a mobile application uses a pervasive device which learners use in everyday life, without any learning curve. The chapter will discuss the development rationale and the features of a mobile application, NIEmGeo, that is specifically geared towards providing a customizable learning platform for collaborative knowledge base development. The application is developed specifically to facilitate and manage field-based authentic learning and provides a platform to integrate information collected from the field, learner's own interpretations of the environment and own inputs, and subsequently to collate all layers of information to create a spatially cognizant knowledge that is customized to provide the intended learning outcome for any specific research.

In a learning structure developed for integration of field-based authentic information, collected both synchronously and asynchronously by groups working separately, Chatterjea (2012) proposed a five-stage learning framework. That framework started with a classroom-based conceptual framework development, spearheaded by the teacher. It went through three stages involving theoretical knowledge dissemination, conceptual understanding and hypothesis building prior to the actual exposure to the outside world. Stage 4 involved actual data collection, when the data collection, data verification, cross-referencing and data collation are backed by the students' prior understanding of the topic of research. This stage also involves data analysis, explanation and knowledge integration, while the last Stage 5 involves generation of a concept map, supported by the integration of all collected, analysed and organized data. The present mobile application works on the Stage 4 of this framework, making data collection and data management integrated through a common mobile platform and subsequently also working on Stage 5, where learners integrate all the learning points to make meaningful understanding of a given concept.

This research used mobile technology as a platform for field-based learning to take advantage of the existing pervasive use of the device by today's learners. The development also utilized the affordances available on mobile devices today to integrate spatial learning. This included features that are deemed essential in any authentic learning environment and may be listed as:

a. Geo-tagging ability, for locational context
b. Visual data sets, for evidence from the field
c. Customized, learner-centred data recording, for flexibility of use in any environment
d. Qualitative inputs, for flexibility of response
e. Collaborative data sets, to reduce individual's workload and yet get enough data for analysis

These goals and the smartphone platforms provided a pervasive, affordable, single-device system that is fairly easy to use. The application, NIEmGeo, was originally developed for iOS platform, but was later extended to Android as well.

Research on Mobile Learning in Authentic Environments

Authentic learning invariably involves some exposure to the real world and hence includes field-based studies. Many researchers have discussed whether integration of IT actually enhances authentic learning in the field. Kaasinen (2003), Pascoe et al. (1999), Baldauf et al. (2007) and many others emphasized the importance of learners being able to author their own findings from the field, the importance of customizing learning specifics. Location-aware services available on mobile phones assist users to access field information. However, such services do not usually cater to the specific needs of individual learner, though there is general information made available to all on the go. The emphasis on authentic learning, however, should point to knowledge creation to suit specific learning goals. Thus, location-aware user authoring is of great importance.

Authentic learning using mobile devices has attracted the attention of many researchers, and the learning exercises have been studied by authors to examine the efficacy of such learning processes. Lim et al. (2006a, b) discussed using mobile phones for doing learner-guided fieldwork, while Chatterjea (2008), Chatterjea et al. (2008), Goh et al. (2006, 2008), Kim et al. (2008), Chang et al. (2012), Nguyen et al. (2008, 2009), Puspitasari et al. (2007), Quach et al. (2010), Razikin et al. (2009), Theng et al. (2006, 2007), Vo et al. (2007) and Vuong et al. (2007) discussed the use of mobile phone applications to conduct collaborative, synchronous field to classroom learning, using geo-referenced information, input of multimedia elements and multilayered tagging as well as grouped fieldwork using maps on servers, to enhance field observation and field-assisted learning. In all these efforts, the common aspect has been the learners' exposure to authentic locations, learners' freedom to classify and record information, gather primary data from the real world, and all these are done using the mobile platform. While researchers have moved from studying the use of various types of IT affordances, along with the advent of new IT equipment, it is evident that the studies focus on the use of changing technologies to gain authentic knowledge and that every study puts emphasis on more and more freedom to learn from the world outside classrooms.

With increasing emphasis on knowledge sharing using technology, the learning scenario has arrived at a stage where the difference between sources of knowledge has almost disappeared. Learners now have the freedom to access information from sources available easily through mobile devices at any place and at any time. With the Web 2.0 tools, the possibility of users to customize, edit, share information from online device has become common (Soon et al. 2008), and as Rost and Holmquist (2010) mention, use of a mobile device to do out of classroom studies satisfies the working style of today's learners. Tarumi et al. (2007) commented that students have difficulty linking authenticity to classroom-based knowledge, but the provision for input of location-aware and customizable information provides a meaningful outcome for students involved in collaborative knowledge development in the field using mobile phones.

One great advantage of using mobile devices (phones and tablets) is the almost universal use of these devices by the learners of today. The learning curve for these devices is non-existent. This makes any learning strategy using these devices readily acceptable to today's learners. In order to incorporate mobile devices effectively in the learning scene of today's learners, Clough et al. (2008) and Patten et al. (2006) emphasized that educationally appropriate application should be built on a combination of collaborative, contextual and constructivist principles' and that mobile devices can particularly support constructivist learning through data collection, site-initiated customizable inputs, location awareness and collaborative work. So far, there is an agreement among researchers that working in the field requires a device that can perform multiple tasks to facilitate learning and that mobile devices do have the advantage by being small, integrative, multitasking and ubiquitous to learners. But there are still areas that need to be addressed, such as a need for an integrative work platform, seamless progression from field to classroom knowledge creation, sharing for knowledge building on a collaborative work and many more.

The present development, NIEmGeo, aimed to close some of these gaps to provide the learner with a tool to perform customized data collection and sharing of location-aware data to enhance authentic knowledge creation.

NIE MGeo Design Considerations to Support Authentic Learning

For an authentic learning exercise to be effective, fieldwork requires a system that is easy to use, integrative, multifeatured to support spatial analysis and able to support constructivist learning, so that the platform used does not come in the way of actual learning. Several features are usually seen as essential, such as:

1. Geo-spatial reference of the field sites: to provide reference for spatial analysis
2. Photographic records of field evidences and features: to enhance the authenticity of the observation
3. Records of large number and variety of quantitative data: to give opportunity to do a real-life problem-solving, rather than customized data set analysis
4. Customizable data recording field: to allow any type of data collection, to suit specific requirements of the study
5. Facility to record qualitative data: to ensure flexibility in observation
6. Facility to record personal reminders and comments for assistance in the field: to help in research management

The above considerations guided the design parameters in the development of mGeo. But it did not stop there. While field exposure gives a real-life experience, any fieldwork supporting authentic learning needs to help the learner to make sense of the spatial and contextual information after completion of the fieldwork. With this objective, MGEO was planned to provide a two-stage support authentic learning. After the field exposure where qualitative and quantitative data are collected, learners proceed to do post-field data sharing, management and analysis. The application also assists in cooperative data development. Finally, after all the learning negotiations, the application extends to reporting the findings, using other available affordances. The post-field operations with the application are as follows:

1. Data upload and data download from a shared location: users can use others' data sets and reduce individual workload without impinging on the limited time.
2. Users can use the platform to discuss and create new data and a shared knowledge base, through comments from all.
3. 3 Data sets are visualized on maps, to provide the much-required geographic information, without researchers having to acquire knowledge in GIS software.
4. The collective data sets are quantitatively organized and analysed, using software such as MS Excel.

The foremost design consideration for this application was to make it user-friendly, with almost no learning curve. Use of smartphone as a platform ensured that the learners do not have to learn the use of new equipment. This encourages more focus on the actual fieldwork, as the user is able to concentrate more on the data collection and observation, rather than moving through different equipment and formats. A smartphone is light, easy to use, and has several built-in features, such as a camera (both still and video), host applications that can measure various parameters, such as heights of objects, angles, distances and other similar useful field applications, has geo-referencing capability that is very accurate, as long as mobile network is available. Added to all these is the almost ubiquitous presence of smartphones in our daily lives—a gadget that is already in everyone's possession. The first development of MGEO was done on iOS platform, as more than 50% students at the university were using iOS. Subsequently, the application was also extended to the Android platform as well.

The specific features incorporated in mGeo are as follows:

1. mGEO is a platform for recording both quantitative and qualitative field data and takes out the need for reliance on pen and paper. This reduces the risks of data loss.
2. Data fields are fully customizable, to suit all types of field investigation. This helps in keeping it open for various types of field investigation, particularly since geography is a multidisciplinary subject and may involve different types of data for different situations.
3. Photographic data form important parts of field evidences for landscape evaluation, process analysis and geographic understanding. The application utilizes mobile phone's built-in system to generate and record such evidences.
4. All data are geo-tagged, to enable geo-spatial mapping, using free software Google Earth, and Google Maps. This takes out the need to use proprietary GIS applications such as ArcGIS, which are expensive, have steep learning curve, require dedicated computer facilities and are not universally available. NIEmGeo thus is equipped to provide spatially referenced locations of field sites, with terrain and other details, for learners to analyse the geographic processes operating there, without using GIS software. It is, therefore, inclusive enough to allow field-based learning for all without additional costs and additional affordances over what is already available.
5. Finally, fieldwork in geography often involves working in remote locations where Internet connections may not be available. Using data transfers and uploads that require Internet connectivity can render the exercise futile or, at best, extremely expensive when roaming data access becomes an essential requirement. mGEO has the ability to record and store all data inputs in the device and upload data to the server when reliable and affordable Wi-fi connections are available.

Working with MGeo: On-Field and off-Field Support

NIE mGeo incorporates a two-tiered geographical analysis system: (i) during fieldwork data recording and (ii) post-fieldwork analysis. The details of workflow are given below:

Fieldwork Phase: Field Observations and Measurements, Field Assessment and Field Decisions

Prior to fieldwork, userids and passwords and specific coloured pins are assigned to each group, so that individual groups' inputs can be easily identified and all data inputs are group specific and safe. Currently, the application can take up to eleven groups, each with different coloured pins. The application requires students to download the application on their mobile devices and login before they go to the field. This loads the field-site map on the mobile device. This is particularly useful for locations that are remote, may not have Internet connectivity or are overseas with potentially high costs of roaming network usage. However, if for some reason this is not done, students can still use the application as usual, although the background map is not available while the work is done. In the field, students create data from their field locations. The following are the steps for using the application in the field.

1. NIE mGeo works on the principle of providing a visual representation of the location/object. This makes the user conscious of the spatial reference and lends relevance to the work done at any given site. Figure 8.1 shows the screen views for data entry:

    The three steps show the sequence followed to create data that is contextual and spatially referenced. The photographic records are geo-referenced, using the location features in the mobile device. Upon completion of the field exercise,

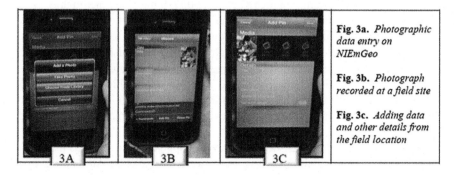

Fig. 3a. *Photographic data entry on NIEmGeo*

Fig. 3b. *Photograph recorded at a field site*

Fig. 3c. *Adding data and other details from the field location*

Fig. 8.1 Steps for field data entry on NIEmGeo

these data are used for mapping work, to provide a spatial reference to the field locations.

2. After the photographic images are recorded, students can input other types of data. NIE mGeo allows quantitative data of any nature, and the data fields are totally customizable, depending on the specific needs of any type of investigation. The teacher/facilitator needs to create the data fields at the back end, prior to the field investigation. The inputs are categorized and can have pull-down menus to reduce field input errors. Due to space constraint, currently six buttons can be accommodated for this data input.
3. However, if research requires more quantitative data types, mGEO offers a very long field for data entry in the next step, allowing the researcher to look at more variables. Here, the application allows the researcher to extend data input by using the non-specific data field. This provides flexibility and makes it user-centric. The data field allows quantitative data, in any preferred order, with no limitations on the number of entries (Fig. 3c) as long as individual inputs are separated by a comma. This format is to organize and tabulate the data later for analysis.
4. Sometimes, quantitative data are not complete without some qualitative inputs. mGEO allows for input of qualitative descriptions, additional information, personal views and comments. All these not only make the data input robust but also provide value-added information that helps the subsequent analysis. For a field analysis, this feature works as the good practice of having a blank space in the data sheet for recording any additional observations that may not be preconceived but might be useful in the field. Field studies often present uncertain and unexpected circumstances. mGEO offers an open solution for such unexpected situations in field studies.
5. In addition to recording data of various nature, the application also allows to record personal comments, reminders, notes of the researcher, which even when the data are uploaded remain in the personal mobile device—only for the researcher to refer to. This works as a personal diary and can be drawn upon if some specific data later appear to present a query. This makes the mGEO a totally pen–and–paper free working environment.
6. The final data input facility in mGEO is the ability of the researcher to draw field sketches, using the tablet or a phone. Sketches can be drawn with choice of four colours, varying widths of lines and input of text for annotation. This feature allows the researcher to record a certain perspective/landscape/process. The annotation can be extended to all images taken and recorded in the mobile device. The images can be reloaded and annotated using this feature and then saved again, with the usual geo-reference. This feature is a newer addition to make the application a truly complete field assistant and to make use of pen and paper redundant.
7. The most defining characteristic of the mobile application, however, is its ability to be used in field situations where no network is available. All data are first saved in the device. This is especially useful in remote field locations, where

either Internet can be erratic or could involve high data roaming charges. When a reliable network connection is available, the data are uploaded by the researcher, using the uploading protocol.

Post-field Phase: Data Analysis and Research

A good analysis of any issue requires not just location-based data, but also a substantial volume of data to yield any worthwhile result. Working in the mGEO environment, many groups can work together, gather data from allocated locations, upload the data to the server and then share the entire database for the final analysis. This gives the researcher the opportunity to examine the problem from a broad base and develop an in-depth understanding, which would have been impossible, if done singularly or in a short time. This feature makes mGEO a very useful tool for field research involving large classes. The screen capture of the uploaded data is shown in Fig. 8.2.

The salient features of this phase of the work are:

1. Once a reliable network is available, the data from individual mobile devices are uploaded by each group, using their own userid and password. Once uploaded, the entire database is visible to all other groups working on the same project. This helps in developing a large database, through the collaborative effort by all individual groups.
2. While the entire data set is visible to all groups, only the authoring group retains the editing access, rest having just 'Read Only' access.

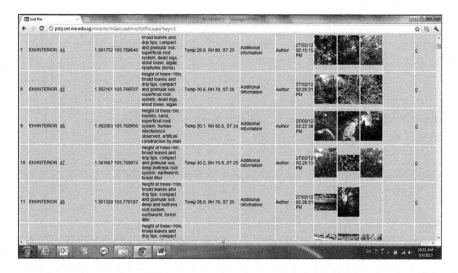

Fig. 8.2 Locational, quantitative and qualitative data uploaded on the server by groups

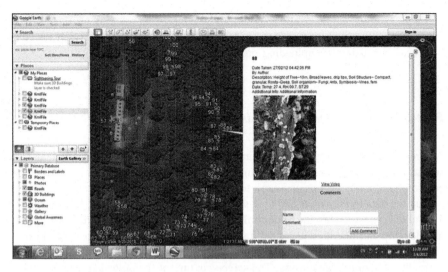

Fig. 8.3 Inset showing data and comment field used by all groups and the lecturer

3. At this stage, all groups as well as the lecturer can query any data or provide comments, which are saved in the server. This way every piece of information can be queried and validated, and in the process, the learners can develop their own knowledge. Figure 8.3 shows a view of such a query, which has all data, as well as the image.

The comments can be added by anyone in the group, while lecturer keeps track of such discussions and can intervene, if necessary. This activity offers the opportunity to the learners to negotiate knowledge building, with adequate scaffolding from the lecturer. This not only allows the lecturer to keep track of students' work, but it also becomes a potential platform for negotiation and collaboration (Fig. 8.3). Peer-reviewed and collaborative development of the geographical understanding, thus, is the focus of the development.

Knowledge-Building Phase: Data Assessment, Data Representation and Synthesis

Once all data are uploaded, negotiated and discussed online, the learners can engage in their own use of the data to assess, analyse and finally develop a synthesized view on the given problem. The first requirement for this analysis is to assess the locational characteristic of the data. NIEmGeo's spatial references are used to generate maps by using Google Earth. The recorded locational references are exported to Google Earth, and maps with the location pins are generated in .kml format, which can be viewed in the browser. Figure 8.4 shows such maps generated

Fig. 8.4 Examples of maps generated from field locational data (the above images could be put together to create one figure)

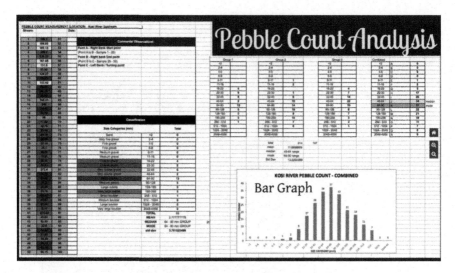

Fig. 8.5 Example of use of graphs and tabulation, directly from mGEO

by students working in groups in different parts of the forest, collecting spatial data. With the terrain information of the field location, learners are now able to draw connections between the specific landscape and the collected data. This allows lateral integration of theoretical and practical knowledge.

The next analytical tool of the application is the affordance of quantitative analysis. The numerical data can be exported as .xls file and tabulated as required, and then, further quantitative analysis can be done in MS Excel (Fig. 8.5). This allows the researcher to use any number of analytical tools, to conduct any relevant type of quantitative analysis.

Besides this, as the main data table with all numerical data, images and any added comments can be accessed by all researchers, the entire data can be used as a relational database for in-depth analysis. When put together, such a set-up assists in synthesizing the information gathered in the field with the learned concepts in the classroom.

Examples of Location-Aware Field Analysis

Students using the application have been involved in creating context-rich data from remote locations and using that data to study physical phenomenon and explain conceptual theories. One group of students conducted fieldwork along Narmada River in Central India to study streamflow variations in the river flowing through a narrow and steep gorge on marble rocks. Another group worked along Kosi River, India, to observe the flow variations on a boulder bed channel. In both the cases, students going across and along the stream measured velocity and other parameters in the stream channels, plotted the isovels, using Google Earth locations, and calculated the flows and drew bar charts, using MS Excel. Figures 12.5 and 12.6 show some of the outputs of these studies done by students, who worked in groups, shared the data on the mGEO platform, used Google Earth and Excel to do the mapping and calculations.

With the features allowing mapping, calculations, tabulations, graphs and charts, ability to fall back on collective comments and solutions, ability to call upon data and field details from other groups and location, to be able to connect numerical data with locational tags, the researcher is equipped with all the necessary tools to perform in-depth analysis. This offers the researcher an integrated platform for work from field to the classroom, with support from not just the peers but also the lecturer. NIEmGeo, can, therefore, work as a platform that supports the notions of learner-centred activity as well as one that integrates scaffolding at the right time and degree.

Details of several field-based research done using the application have been described and explained in Chatterjea (2008). These showcase the various affordances of mGEO that facilitate making sense of real-life situations and support student-centred work, while not comprising on quality of research stemming from limited data. In considering the ability to accommodate multiple group inputs and allowing for data sharing, mGEO optimizes the limited time available in practical school environment and yet provides the opportunity to do in-depth investigation.

Data Security Features of NIEmGeo

As mGEO is an application that allows multiple users to input data for collective usage later, data security is an issue that needs to be taken care of. Certain features are, therefore, in place to ensure tamper-proof usage.

1. Only the registered logged in user in a group can upload data or edit the group's upload. This prevents the data from getting tampered by users in other groups.
2. The application detects duplication of observation station identity and allows only discrete station numbers throughout the entire data set. This prevents individual data to be confused with another and maintains data integrity.
3. There is a security protocol which only keeps the latest image and data, so that the databases are not duplicated or corrupted.
4. Even after the upload, all data still remain in the mobile device and any further updates for the same location replace the older data on the server, to reduce confusion. Use of the in-phone storage, choice of data upload time, as well as subsequent uploading to update data are aimed at covering the need for safe data storage, and data editing and updating as required by the researcher.

Usefulness in the Classroom Situation

Since the application can display 11 colours, the class can be divided into as many groups, to facilitate data collection in huge volumes and yet allocate smaller chunks of work for each group. This is particularly aimed at empowering teachers with severe time constraints to still engage the students in authentic research work and provide opportunities of exposure to real-life field situations, even when time allocation for such activities is limited. Each group can do just a part of the work. But with facilities of data sharing, the research phase can still take advantage of large databases collectively developed by the entire class.

Maps, with data pins of each group, can be generated. As different groups use different pin colours, individual group's work can be seen and assessed by the lecturer. This makes assessment of students' work objective and meaningful.

As each point of observation is backed by data sets that can be queried, the teacher can check the quality of the work and at the same time provide targeted assistance and guidance. This allows students the freedom to work independently, while being guided by the teacher.

Figure 12.4 showcases such group work in a class of 80 students, where 10 groups of students worked in 10 different parts of a dense tropical rainforest, collecting various types of environmental data, simultaneously. Later, these data were used to draw up various isolines for analysing the environmental conditions in the remote forest interior and comparing it with the external conditions (Chatterjea 2014). This research showing the impact of fragmentation of the forest and forest boundary environment was possible using the various affordances of NIEmGeo, to conduct shared field data collection simultaneously, and using the collective data to perform in-depth analysis. One of the resulting isovel maps is shown in Fig. 8.6. Details of this work are discussed in Chatterjea (2010).

The ability to export numerical data to spreadsheets allows students to perform data analysis easily, as well as perform graphing work, as shown in Fig. 8.5. When all the various features of the application are used, the students can cover the Stages 4 and 5 of the framework proposed by Chatterjea et al. (2008) and Chatterjea (2008), where the conceptual framework introduced in class in Stages 1 to 3 can be further extended to authentic locations where the hypothesis can be tested, and new knowledge can be created, based on student-led investigations and analysis (Stage 4). Finally, as the entire sets of observations are shared, the knowledge derived becomes a negotiated knowledge and thus more meaningful to the learners (Stage 5). Thus, instead of standing alone as a technological intervention in learning, mGEO provides a broad-based platform, to integrate the various stages of learning with a definite learning goal that incorporates both scaffolded guidance and learner-centric knowledge development.

C2	Start Point	Dist LB	Dist RB	Depth	Streamflow 1 (no of revolutions in 10s)		Streamflow 2 (no of revolutions in 10s)		Comments
					0.5m	1.0m	0.5m	1.0m	
Station 1	LB	0m	46m	1.50m	63 10/13	11/14/40	5	14	
Station 2		3m	43m	13.30m	13	6	No Data	No Data	
Station 3		15m	31m	13.37m	29	9	7	19	
Station 4		18m	28m	15.90m	No Data	No Data	No Data	No Data	
Station 5		26m	20m	18.94m	14	13	10	9	

C3	Start Point	Dist LB	Dist RB	Depth	Streamflow 1 (no of revolutions in 10s)		Streamflow 2 (no of revolutions in 10s)		Comments
					0.5m	1.0m	0.5m	1.0m	
Station 1	RB	75m	0m	8.73m	32	27	No Data	No Data	
Station 2		56m	9m	9.83m	18	22	75	92	
Station 3		50m	25m	20.70m	68	77	No Data	No Data	
Station 4		35m	40m	22.40m	121	23	99	4	
Station 5		21m	54m	13.60m	No Data	No Data	No Data	No Data	
Station 6		0m	75m	2.37m	5	9	4	4	

C4	Start Point	Dist LB	Dist RB	Depth	Streamflow 1 (no of revolutions in 10s)		Streamflow 2 (no of revolutions in 10s)		Comments
					0.5m	1.0m	0.5m	1.0m	
Station 1	LB	0m	73m	2.17m	4	6	0	1	
Station 2		15m	58m	10.33m	8	54	37	54	
Station 3		21m	52m	14.47m	12	97	9	0	
Station 4		33m	40m	20.50m	25	36	No Data	No Data	

C5	Start Point	Dist LB	Dist RB	Depth	Streamflow 1 (no of revolutions in 10s)		Streamflow 2 (no of revolutions in 10s)		Comments
					0.5m	1.0m	0.5m	1.0m	
Station 1	RB	73m	0m	20.50m	25	36	0	1	Same as C4 Station 4
Station 2		59m	14m	15.57m	10	55	46	19	Streamflow 1- No ear
Station 3		43m	30m	20.47m	4	11	34	55	

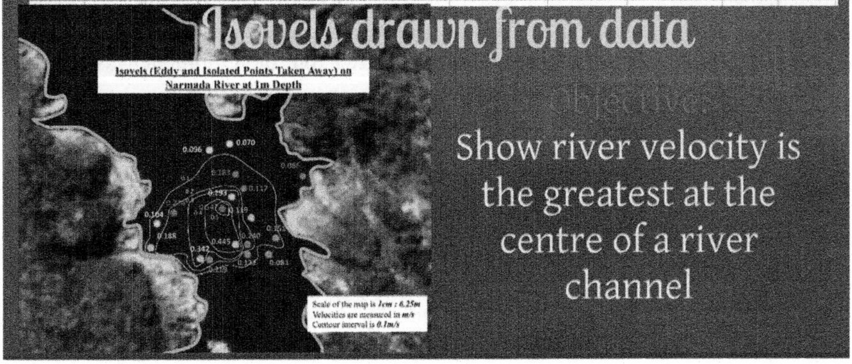

Fig. 8.6 Map-based data analysis from the collective database

Learning Experiences of Students Using NIEmGeo

The application has so far been used by more than 2700 students and teachers in primary and secondary schools in Singapore, to conduct field-based studies in geography, social studies and science. The objective in all these was to incorporate IT-infused experiential learning in the regular classroom exercises.

While detailed data on the usage patterns of students in schools were not recorded by the teachers, usage patterns of trainee teachers at the National Institute of Education are discussed in Chatterjea (2010) and the following can be summarized:

In the fieldwork stage,

1. The learning curve for the application was very flat, with all students being very comfortable in the use of the application.
2. There was complete data security at all stages of the field investigations.
3. Recording of geo-referenced data was error-free even in remote field locations.
4. Uploading of data was smooth with Wi-fi connection, subsequent to return in Wi-fi environment.

5. Huge numbers of geo-referenced locational data were uploaded by all groups.
6. Students generated huge amounts of photographic data which were shared on the server.
7. Uploading of both quantitative and qualitative data input was equally heavy.

In the data analysis stage,

1. Everyone used mGEO as a data repository and accessed the database for their analysis.
2. While use of other group's data and images was evident in their analysis, students tended to use their own data and images more heavily.
3. All used mGEO to generate maps on Google Earth and generate graphs using MS Excel.
4. Graphs and maps were used together to arrive at conclusions.
5. Use of the built-in facility of negotiated knowledge development was not universal, though there was some activity among some groups.
6. Some students used the sketching and annotation facilities and incorporated it in their reports.

Overall, students as well as the teachers using the application gave positive feedback about the usefulness of the application in making field-based learning a meaningful exercise.

Conclusion

Experiential and inquiry-based learning through fieldwork has become an essential element in many disciplines. But issue of allowing all students to engage in fieldwork is always fraught with problems such as time constraints, difficulty to get affordable network connections and expensive gadgets, and a general lack of a seamless system that can serve the fieldwork and also provide a workable platform for further post-field analysis, so that the exercise has some in-depth learning outcome. mGEO was set to solve some of the nagging problems in fieldwork in geography courses: too many cumbersome equipment, the problem of having to record data on paper while juggling with multiple equipment and then having to safeguard this piece of paper from getting lost. It also aimed at providing geo-referenced data and provides a workable solution to the common woes of a teacher trying to juggle between limited curriculum time, need for relevant and useful field experience and data collection that would be useful for actual analysis, without impinging on students' out of college time. The application provided these effectively, judging from the usage patterns of more than 3000 geography students at various levels of study, as mentioned earlier. The issue of using maps for a spatial analysis, using big database for meaningful analysis, and cooperative database development through academic negotiations are some of the features that make pedagogically sound learning environments and make connections between the classroom and the outside world. Students on the task generally responded favourably on the usability of the application, with most mentioning that it helped them to visually place their data locations on the map and recording each data point details. The post-field analysis of students' work using the substantial data set was clearly an advantage for all the groups.

Another enhancement of the latest version of mGEO is the affordance of being able to do field sketching on the same mobile device, in the same mGEO environment, without having to switch to another application or device to do that. This facility not just allows students to record field features but also forces them to observe salient features and analyse the relationships between forms and processes operating at the location. No geography field tool is complete without this facility to do field sketching.

Indeed, authentic learning through fieldwork can be supported by applications on mobile phones to not just help in field observations but also in harnessing mobile technology to manage the total learning environment. This includes support for field data collection to post-field data organization and analysis, which allows students to complete the whole learning circle and create a knowledge base on the intended topic. NIEmGEO has allowed large groups of students to be engaged in doing authentic field-based study and do in-depth analysis through collective database development and socially negotiated knowledge.

What makes mGEO a very useful tool both in the field and back in class is its multiple features that attempt to provide a complete field and research assistance, while also incorporating provisions for collaborative work and negotiated knowledge development, while being in full support of the teacher, making it a truly authentic experience.

Acknowledgements This is an outcome from the research project SUG 21/11 KC. Thanks are due to Ms Er Ching Wei Eveleen of Centre for E-Learning, NIE, for writing the codes for the application, following all instructions strictly and complying to the varied instructions on the features and workflow of the application development.

References

Baldauf, M., Dustdar, S., & Rosenberg, F. (2007). A survey on context-aware systems. *International Journal of Ad Hoc and Ubiquitous Computing, 2*(4), 263–277.
Chang, C. H., Chatterjea, K., Goh, D. H. L., Theng, Y. L., Lim, E. P., Sun, A., et al. (2012). Lessons from learner experiences in a field-based inquiry in geography using mobile devices. *International Research in Geographical and Environmental Education, 21*(1), 41–58.
Chatterjea, K. (2008). Use of virtual workplace as a new learning environment: a case from Singapore. *Handbook of research on virtual workplaces and the new nature of business practices* (pp. 301–316).
Chatterjea, K. (2010). Using concept maps to integrate hierarchical geographical concepts for holistic understanding. *Research in Geographical Education, 12*(1), 21–40.
Chatterjea, K. (2012). Use of mobile devices for spatially-cognizant and collaborative fieldwork in geography. *Review of International Geographical Education Online, 2*(3).

Chatterjea, K. (2014). Edge effects and exterior influences on Bukit Timah forest, Singapore. *European Journal of Geography, 5*(1).

Chatterjea, K., Chang, C. H., Lim, E. P., Zhang, J., Theng, Y. L., & Go, D. H. L. (2008). Supporting holistic understanding of geographical problems: Fieldwork and G-Portal. *International Research in Geographical and Environmental Education, 17*(4), 330–343.

Clough, G., Jones, A. C., McAndrew, P., & Scanlon, E. (2008). Informal learning with PDAs and smartphones. *Journal of Computer Assisted learning, 24*(5), 359–371.

Goh, D. L. H., Theng, Y. L., Lim, E. P., Zhang, J., Chang, C. H., & Chatterjea, K. (2006). G-Portal: A platform for learning geography. *Encyclopaedia of Portal Technology and Applications*, 547–553.

Goh, D. L. H., Theng, Y. L., Lim, E. P., Zhang, J., Chang, C. H., & Chatterjea, K. (2008). Learning geography with the G-portal digital library. In P. Rees, L. MacKay, D. Martin, & H. Durham (Eds.), *E-learning for geographers: online materials, resources, and repositories* (pp. 260–269). Hershey, PA: Idea Group Inc.

Kim, T. N. Q., Razikin, K., Goh, D. H. L., Nguyen, Q. M., Theng, Y. L., Lim, E. P., et al. (2008, December). MobiTOP: Accessing hierarchically organized georeferenced multimedia annotations. In *International Conference on Asian Digital Libraries* (pp. 412–413). Berlin, Heidelberg: Springer.

Lim, E. P., Wang, Z., Sadeli, D., Li, Y., Chang, C. H., Chatterjea, K., et al. (2006, November). Integration of Wikipedia and a geography digital library. In *International Conference on Asian Digital Libraries* (pp. 449–458). Berlin, Heidelberg: Springer.

Lim, K., Hedberg, J., & Chatterjea, K. (2004). Pictures in place: how teenagers use multimedia messaging to negotiate, construct, and share meaning about geographical tasks. In *EdMedia: World Conference on Educational Media and Technology* (Vol. 2004, No. 1, pp. 1179–1186).

Lim, E. P., Zhang, Jun, Yuanyuan, Li, Wang, Zhe, Chang, C. H., Chatterjea, K., et al. (2006b). G-Portal: A cross disciplinary digital library research program from Singapore. *IEEE Technical Committee on Digital Libraries Bulletin—Special Issue on Asian Digital Libraries, 3*(1), 20–56.

Nguyen, Q. M., Kim, T. N., Goh, D. H., Lim, E. P., Theng, Y. L., Chatterjea, K., et al. (2009, May). Sharing hierarchical mobile multimedia content using the MobiTOP System. In P. Kellenberger (Ed.), *Mobile data management 2009* (pp. 637–642). Institute of Electrical and Electronics: IEEE Computer Society.

Nguyen, Q. M., Kim, T. M. Q., Goh, D. H. L., Theng, Y. L., Lim, E. P., Sun, A., et al. (2008, September). TagNSearch: Searching and navigating georeferenced collection of photographs. In B. Christensen-Dalsgaard, D. Castelli, B.A. Jurik, & J. Lippincott (Eds.), *Research and Advanced Technology for Digital Libraries: Proceedings 12th European Conference, ECDL 2008, Aarhus, Denmark, September 14–19, 2008* (pp. 62–73). Berlin: Springer. On 10/5/2012.

Pascoe, J., Ryan, N., & Morse, D. (1999, September). Issues in developing context-aware computing. In *International Symposium on Handheld and Ubiquitous Computing* (pp. 208–221). Berlin, Heidelberg: Springer.

Patten, B., Sánchez, I. A., & Tangney, B. (2006). Designing collaborative, constructionist and contextual applications for handheld devices. *Computers & Education, 46*(3), 294–308.

Puspitasari, F., Lim, E. P., Goh, D. H. L., Chang, C. H., Zhang, J., Sun, A., et al. (2007). Social navigation in digital libraries by bookmarking. In D.H.L. Goh, T.R. Cao, I.T. Slvberg, & E. Rasmussen (Eds.), *Asian digital libraries. Looking Back 10 years and forging new fontiers* (pp. 297–306). Berlin: Springer.

Quach, H. N. H., Razikin, K., Goh, D. H. L., Kim, T. N. Q., Pham, T. P., Theng, Y. L., et al. (2010, August). Investigating perceptions of a location-based annotation system. In A. An, P. Lingras, S. Petty, & R. Huang (Eds.), *Active Media Technology: Lecture Notes in Computer Science, 6th International Conference, AMT 2010, Toronto, Canada, August 28–30, 2010. Proceedings* (pp. 232–242). New York: Springer.

Razikin, K., Goh, D. H., Theng, Y. L., Nguyen, Q. M., Kim, T. N., Lim, E. P., et al. (2009, October). Sharing mobile multimedia annotations to support Inquiry-based learning using MobiTop. In D. Hutchinson et al. (Eds.), *Proceedings Active Media Technology: 5th International Conference* (pp. 22–24). New York: Springer.

Rost, M., & Holmquist, L. E. (2010). Tools for students doing mobile fieldwork. In T. Goh (Ed.), *Multiplatform E-learning systems and technologies: Mobile devices for ubiquitous ICT-Based education* (pp. 215–228). Retrieved from https://doi.org/10.4018/978-1-60566-703-4.ch013.

Soon, C. J., Roe, P., & Tjondronegoro, D. (2008, March). An approach to mobile collaborative mapping. In *Proceedings Symposium on Applied Computing* (pp 1929–1934), Fortaleza, Ceara, Brazil.

Tarumi, H., Satake, F., & Kusunoki, F. (2007, July). Collaborative learning with fieldwork linked with knowledge in the classroom. In *IADIS International Conference, Mobile Learning 2007*, Lisbon, Portugal.

Theng, Y. L., Tan, K. L., Lim, E. P., Zhang, J., Goh, D. H. L., Chatterjea, K., et al. (2007, June). Mobile G-Portal supporting collaborative sharing and learning in geography fieldwork: An empirical study. In E. Rasmussen, R. R. Larson, E. Toms, & S. Sugimoto (Eds.), *Proceedings of the 7th ACM/IEEE Joint Conference on Digital Libraries* (pp. 462–471). New York: Association for Computing Machinery.

Theng, Y. L., Li, Y., Lim, E. P., Wang, Z., Goh, D. H. L., Chang, C. H., et al. (2006, November). Understanding user perceptions on usefulness and usability of an integrated Wiki-G-portal. In S. Sugimoto, J. Hunter, A. Rauber, & A. Morishima (Eds.), *Digital Libraries: Achievements, Challenges and Opportunities Proceedings of 9th International Conference on Asian Digital Libraries, ICADL 2006, Kyoto, Japan*, (pp. 507–510). Berlin: Springer.

Vo, M. C., Puspitasari, F., Lim, E. P., Chang, C. H., Theng, Y. L., Goh, D. H. L., et al. (2007, June). Mobile digital libraries for Geography Education. In E. Rasmussen, R. R. Larson, E. Toms, & S. Sugimoto (Eds.), *Proceedings of the 7th ACM/IEEE Joint Conference on Digital Libraries* (pp. 511–521). New York: Association for Computing Machinery.

Vuong, B. Q., Lim, E. P., Sun, A. Chang, C. H., Chatterjea, K., Goh, D. H. L., et al. (2007, September). Key Element-Context Model: An Approach To Efficient Web Metadata Maintenance. In N.F. Kovacs, & C. Meghini (Eds.), *Research and advanced technology for digital libraries* (pp. 63–74). Berlin: Springer.

Kalyani Chatterjea is an avid advocate of physical geography, more specifically geomorphology. Her research focuses on urban geomorphology, rainforest hydrology, environmental change and channel response. Her special interest includes music and painting and a good holiday destination where one can enjoy wildlife and/or mountains.

Chapter 9
Location-Aware, Context-Rich Field Data Recording, Using Mobile Devices for Field-Based Learning in Geography

Muhammad Faisal Bin Aman and Boon Kiat Tay

Abstract The chapter focuses on the use of mGeo app (see Chap. 8) as a mobile technology tool to identify and mark sites of geographical investigation. These sites are also embedded with data and images in order to consolidate information and to mirror onsite environment conditions. The study of soil compaction and soil permeability along Bukit Timah Nature Reserve trails provides the basis of the geographical investigation. The aim of the geographical investigation is to establish the impact of walking along the designated trails and the building of concrete steps along the trail. The use of NIE mGeo provides a platform whereby students can immediately tag sites of investigation in real time with information and data that they have collected. It also provides a databank for future investigation by other students. The use of NIE mGeo can be replicated by other schools, and they can add on to the data collected, therefore allowing changes and trends of soil compaction and soil permeability to be analysed along the trails in Bukit Timah Nature Reserve for longitudinal studies. The consolidation of data by students will allow a form of active contribution to the ongoing learning process in geographical and scientific inquiry into the study of soil quality. The chapter will primarily discuss (i) the usage and representation of data on mobile devices in situ, (ii) collaborative inquiry processes in situ and the (iii) the role and pedagogical responsibility of the facilitator in the collaborative processes.

M. F. B. Aman (✉)
National Institute of Education, Nanyang Technological University, Singapore, Singapore
e-mail: faisal.aman@nie.edu.sg

B. K. Tay (✉)
Marsiling Secondary School, Singapore, Singapore
e-mail: tay_boon_kiat@moe.edu.sg

© Springer Nature Singapore Pte Ltd. 2018
C.-H. Chang et al. (eds.), *Learning Geography Beyond the Traditional Classroom*,
https://doi.org/10.1007/978-981-10-8705-9_9

Background

The use of mobile technologies to identify and mark sites of investigation in geography learning started with a research project on soil compaction and soil permeability. The project began when students at a public school in Singapore expressed interest in participating in the "Little Green Dot Student Research Grant" organised by a non-governmental organisation—the Nature Society, Singapore. The secondary school students in this project were interested to know how human activities such as forest trekking have an impact on the natural environment, particularly on the soil chemical composition, compaction and porosity. The teachers (both authors in this chapter) were tasked with facilitating and mentoring the students to carry out their field investigation.

The students first identified a site of study—the Bukit Timah Nature Reserve. The teachers then sought advice from two experts from the National Institute of Education (NIE), Nanyang Technological University, Dr. Shawn Lum a Biologist and Dr. Chatterjea, a Geographer. Based on the interactions with and guidance from the two experts, the students were able to design their investigation. Although the students knew what to do in the field and what Data to collect, there was a problem on how they could identify and keep track of the sites of investigation with the available data gathered. Figure 9.1 shows a map of the site and the extent of the study area.

The authors shared this challenge with Dr. Chatterjea, and she suggested mGeo, a mobile app developed by NIE. It provided a platform where students were able to tag information with geographical locations of the sites of investigation. This spatial organization and representation of data helped consolidate their findings, enabling them to analyse the data from a larger geographical perspective. At the time of writing this chapter, the students' research into soil compaction and soil permeability at the Bukit Timah Nature Reserve was ongoing; the focus of this chapter will thus be on developing a critical narrative on the use of mobile technology for students' geographical field investigation. Upon reflection, the use of mGEO offered opportunities for collaboration between other schools' students and has potential for a longitudinal study to be carried out at the Bukit Timah Nature Reserve.

Theoretical Framing

In order to describe the learning activity and to provide a critical narrative of how the study would help students learn better, a constructivist theory of learning (Piaget 1928) was chosen to describe the student researchers as key participants in their learning. This is in contrast to paradigms of learning that consider students as blank slates waiting for information to be written on (Creedy et al. 1992). A constructivist

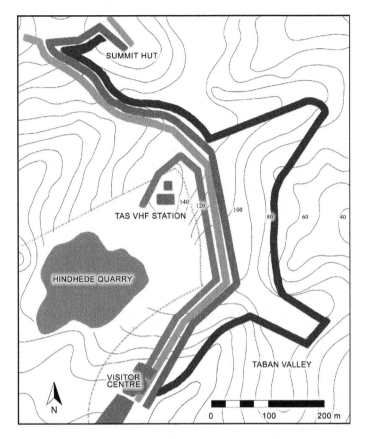

Fig. 9.1 Location map of the Bukit Timah Nature Reserve and the four nature routes

perspective emphasises the agency of a student in making meaning of the information that he/she encounters, thereby resulting in new ways of thinking about the information. Students actively construct knowledge based on their observations and prior knowledge (Mayer 2003), with the intent to interpret the environment and contextualise the content learnt (Peterson et al. 1996). In addition, the sociocultural theory of learning (Vygotsky 1962) describes students as participants who deliberate, reflect and inform one another to build new knowledge. Such collaborative interactions are focused on a specific issue or problem and are sometimes mediated and cognitively guided by facilitators at strategic moments.

In considering both approaches to meaningful learning, inquiry-based learning fits well within the two learning theories reviewed, as student researchers plan, collect data and articulate ideas related to soil porosity, permeability and compaction. Inquiry-based learning therefore allows "learners to get in touch with authentic situations…that are [analogous] to real life [settings]" (Feletti 1993; Li and Lim 2008). Rather than transmitting knowledge from teachers to students in a theoretical and didactic manner, Creedy et al. (1992) believe that inquiry-based learning encourages students to be active learners rather than passive recipients of knowledge.

However, learning geography in a real-life setting often requires management of real-life geographical data. One key challenge for all geographical inquiry is in managing data with location information. In this case, information and communication technology can provide solutions to support students' geographical learning. Indeed, Chang et al. (2012) have suggested that technology can play an important role in geographical inquiry in the field. To this end, the use of software and mobile applications needs to be delineated systematically. Some software applications allow collection of geographical data with a view to explain the topic and theme (Yeh et al. 2006; Wentzel et al. 2005), while others support the students' strategies and skills in obtaining data (Loh et al. 2001). Yet, other software applications allow the creation and sharing of data for analysis and learning (Spigol and Milrad 2008). A key feature of mobile technologies is the spatial freedom to learn (Sharples et al. 2005). The student researcher is therefore not confined or rooted to a specific site.

Literature Review

In addition to the a few studies in Singapore on how field-based inquiry in geography using mobile devices can be carried out (Chang et al. 2012; Nguyen et al. 2008; Chatterjea et al. 2008), studies involving mobile devices and smartphones in the field setting by amateur naturalists have also been previously recorded. The Royal Society for the Protection of Birds (RSPB) in the UK started a 'bird track' for observations and migration patterns. Creet et al. 2005 summarise studies involving the conservation of birds at national, regional and local scales by Corlett et al. (2004), and Vayoula and Sharples (2003). Dion et al. (2007) reveal that in the urban environment in Singapore, MobiTOP, a mobile tagging application could allow people with disabilities share up-to-date accessibility information about buildings and physical structures to assist them navigate in their environment (Razkin et al. 2009). However, the studies in Singapore focused on using an explicit framework for inquiry that involves:

> Creating a need to know;
> Using data;
> Making sense of data, and;
> Reflecting on the learning.

Indeed, this cycle of inquiry draws on the idea advanced by Roberts (2003). The use of technologies can support every stage of the Roberts' (2003) cycle of inquiry, but the affordances of technology support stages 2 through 4 specifically in the field context. Authentic tasks such as the measurement of porosity and permeability of walking trails involve a meaningful investigation and tasks that mirror the work of geographers and earth scientists. In addition, the use of mobile technologies in geography field inquiry provides timely and informative feedback through collaboration with other students through the application (Chang et al. 2012). Students are able to develop practical skills such as observation and recording data on the move in the field. While the G-Portal (Chatterjea et al. 2008) and MobiTOP (Razkin et al. 2009) developments provided some key features that would support geographical inquiry in the field, Dr. Chatterjea from NIE who is one of the researchers in these two earlier projects has gone on to develop the NIE mGEO (mGEO).

The NIE mGEO is a computer-supported collaborative learning (CSCL) that offers the possibility of successful collaborative learning. Educators from other schools could assist and add on to the information, while students are still conducting the fieldwork. Such collaboration requires coordination, structured processes and clearly delineated outcomes. Zurita and Nussbaum (2004) argue that there must be synchronisation and clarity in the roles and tasks of students and educators from other schools in order for field inquiry to be meaningful. What is critical and interesting are the interactions between student researchers, tools and their settings in this inquiry-based learning journey.

While the literature review has uncovered ideas that converge on the supportive role of mobile technology in geographical field-based inquiry, this chapter will present a critical narrative based on actual experiences gained from running the fieldwork using mGEO with the students. The reflections are constructed based on observations by the authors as well as the analysis of students learning artefacts.

Field Investigation Supported by MGEO

After discussing the research problem with the teachers and the experts, the students began by choosing a section of "Route 4" as the route for investigation. Figure 9.2 shows the tagged stations along the chosen route by students.

Soil compaction was measured using a penetrometer, and soil infiltration rate was measured using PVC cylinders and stopwatch (see Figs. 9.3, 9.4 and 9.5). Soil samples were also collected and tested for soil properties. The PVC cylinders were filled up to the mark of 5 cm to measure infiltration rate for different stations along Route 3. Stations were set up along the trail. They consisted of boardwalks, dirt trails and steps made up of rocks and dirt.

At every station, students took two readings each, one on the designated trail where people walked and a parallel area along the designated trail where possible. Due to safety concerns, some stations only have readings collected on the designated trail itself. This is because some of the stations outside the designated trails were on steep slopes or had a lot of debris, making it dangerous for data collection of soil compaction and soil infiltration.

Real-time data collected at each station was recorded, and its location was tagged using the NIE mGEO. The data was uploaded to the server through the

Fig. 9.2 Curved section of Green Route (fourth line in Fig. 9.1) with tagged information represented on mGeo

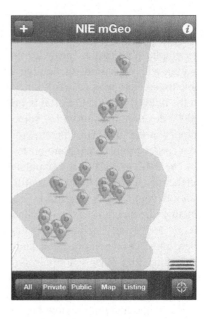

Fig. 9.3 Penetrometer

Fig. 9.4 Penetrometer pushed into soil

Fig. 9.5 Infiltration rate measured

application, with descriptions of the conditions of the stations. Using a tablet (iPad in this case), students also took pictures at the stations to provide a visual image of the condition.

Findings and Discussion

Consolidation and Storing of Data

There is potential in the use of NIE mGEO to collect and store data. The use of the iPad was primarily utilised to capture images, whereas the use of the smartphones was to tag and mark stations. Currently, data, the mGEO, and the platform where mGEO is hosted are accessible and provided free of charge. This should remain. However, the modification to available and past data should be restricted by an administrator so that there is a way to manage the integrity of the data. Nonetheless, mGEO has allowed students to consolidate and upload images and data at specific sites immediately and or at any time. Interestingly, such convenience has highly encouraged students to be meticulous in data collection as they were observed to repeat their readings to ensure accuracy. This has led students to be independent and confident in their use of mGEO and data collected. Importantly, mGEO enables students to revisit tagged sites a few months later, thus reflecting more accurate readings at their selected stations.

Analysis and Making Sense of Data

The teachers opined that in general, the process of collecting, analysing and presenting data with the use of mGEO did support the learning of students. The use of technology not only provides convenience to learners but strengthens the inquiry process and allows students to reflect their procedural and geographical knowledge as well as to consider other factors that may affect outcomes and readings. In other words, rather than being fixated on finding the correct answer, the use of technology in this case shifts the focus of obtaining the right and alternate methods from being fixated to find an answer. However, the learning for students and teachers was not often smooth sailing.

As the project began, it was apparent that the role of the teacher as a facilitator becomes more important, especially utilising technology-enabled devices. It was observed that in the phase 1 of the project, students were too familiar of obtaining immediate feedback and clarification from teachers. Teachers in turn were also accustomed to answer questions too quickly. Instead of allowing students to reflect and think through, in the interest of time, teachers gave in and provided affirmation and answers too quickly. Secondly, it was observed that students did give inputs to one another albeit fleetingly and in a personal manner. Upon closer observations, students' questions and answers to one another seemed to be close ended or non-discursive. Hence, both teachers and students have initially found it uncomfortable and challenging to utilise the mGEO device and also adopt an inquiry process at the beginning of the project. However, it was a steep learning curve for students.

Teachers should not assume that mobile technologies will be able to guide students and offload teachers' responsibilities. On the contrary, the teachers' role as facilitators becomes critical as the students seek avenues to clarify and work on their inquiries. In phase 2 of the project, students revisited the sites along Route 3 with the help of markers on the annotated map. There was also a conscious effort taken by the teachers to redirect questions back to the students and co-construct new information amongst themselves. Rather than directing questions that were of factual, close ended and hypothetical in nature, teachers and students discussed on the "why" and "how" questions. These discussions revolved around the data that were spatially represented at the mobile app map. The data collected from the use of technology provided opportunities for student-to-student discussions as well as student-to-teacher discussions. Hence, understanding and discussing data was central in their geographical inquiry.

Reflecting on Learning

As a recipient of the research grant, students were required to produce a written academic report and present their findings at an environment exhibition. The platform could be perceived as a learning opportunity for students to defend their findings and justify whether their research had met the objectives of the project. The formal experience adds a realistic dimension to their learning. As a team, students had to conduct a project debrief to prepare for the exhibition. It also provides a chance for the authors to understand their analysis and listen to the students' performance. This is in line with the last stage of the geographical enquiry cycle where the teacher should try "to understand what has gone on in the students' minds, what sense they have made of what they studied and whether they have had to rethink what they knew before" (Roberts 2010).

The authors had also reflected on their learning. They believed that teacher training in terms of familiarisation and literacy of the software application should be conducted to improve ICT competencies in any future projects. While the two authors have become very familiar with the technology in this activity, it was due to a prolonged collaboration with the expert at NIE who developed this software application. Compared to the phase 1 of the project, the phase 2 exemplified the ease and efficient use of the mGEO features such as tagging, taking photos and filling information. Students seemed more confident and focused in time. However, the teachers foresee that operational issues such as familiarity with software may affect the motivation of future students if no introductory or preparatory lessons are planned. In retrospect, it would be useful for both teachers and students to undergo workshops together to shorten and quicken the learning curve of utilising location-aware and context-rich data recording devices on site. More importantly, lessons on facilitation and communication skills could improve the inquiry approach and process before going into the field for geographical inquiry.

In general, the use of technology in this study can be described by Roberts' (2003) cycle of inquiry but with qualifications. The inquiry process requires time for teachers and students to gain familiarity and confidence. It was observed that there was an increase in frequency of short and in-depth discussions. Apart from analysing data that were collected immediately at the sites, students were also able to reflect on their annotations and readings in the classroom days later. The technology was particularly helpful for team members who were not present at the sites to further question and understand the data analysis. Students had to feed-forward information and analysis, 'backtrack' and reflect on the processes. Based on what was observed in Phase 2, it could then be argued that there were "mini-cycles of inquiry" that oscillated from students to teachers and amongst students themselves. Indeed, the data and finding from each inquiry cycle provides information to create a need to know in the subsequent inquiry cycles. In fact, there was an absence of a neat delineation of the inquiry processes and cycle proposed by Roberts (2003). From the authors' perspective, a reiterative process that goes back and forth and one that cuts across the different stages at different times would reflect the reality and operation of Robert's (2003) inquiry cycle.

Conclusion

While the experiences in this learning activity have corroborated some key ideas uncovered through the literature review, it has affirmed for the authors the importance of deliberate and careful planning before we bring students out to the field for geographical inquiry. The consolidation of data by students has allowed a form of active contribution to the ongoing learning process in geographical and scientific inquiry into the study of soil quality. The chapter has discussed (i) the usage and representation of data on mobile devices in situ, (ii) collaborative inquiry processes in situ and the (iii) facilitator's role and pedagogical responsibility in the collaborative processes. Mobile technologies together with cloud-based solutions of storing and managing data, and in this case, spatial and environmental data, have certainly provided exciting opportunities for learning geography. However, the authors remain mindful of the need to be well trained to use the technology and to shift the focus of attention to facilitating learning rather than directing learning. The key idea behind the Roberts' (2003) inquiry model is aligned to the constructivist approach of getting students to make meaning of the information they encounter. The teachers' role becomes even more important in ensuring that students are supported in this construction of knowledge, especially when mobile technology use for teaching and learning is becoming ubiquitous.

References

Chatterjea, K., Chang, C. H., Lim, E. P., Theng, Y. L., Goh, D. H., & Zhang, J. (2008). Supporting holistic understanding of geographical problems: Field work and G-Portal. *International Research in Geographical and Environmental Education, 17*(4), 330–343.

Chang, C. H., Chatterjea, K., Goh, D. H., Theng, Y. L., Lim, E., Sun, A., et al. (2012). Lessons from learner experiences in a field-based inquiry in geography using mobile devices. *International Research in Geographical and Environmental Education, 21*(1), 41–58.

Corlett, D., Sharples, M., Chan, T., & Bull, S. (2004). A mobile learning organiser for university students. In *Wireless and Mobile Technologies in Education, 2004. Proceedings. The 2nd IEEE International Workshop on* (pp. 35–42). IEEE. Chicago.

Creedy, D., Horsfall, J., & Hand, B. (1992). Problem-based learning in nurse education: an Australian view. *Journal of Advanced Nursing, 17*, 727–733. Dion, H.G., Sepoetro.

Dion, H. G., Sepoetro, L. L., Qi, M., Ramakhrisnan, R., Theng, Y.L., Puspitasari, F., et al. (2007). *Mobile tagging and accessibility information sharing using a geospatial digital library*. Asian Digital Libraries. Looking Back 10 years and Forging New Frontiers. Series Lecture Notes in Computer Science (Vol. 4822, 287–296).

Feletti, G. (1993). Inquiry based and problem based learning: how similar are these approaches to nursing and medical education? *Higher Education Research and Development, 12*(2), 143–156.

Li, D. D., & Lim, C. P. (2008). Scaffolding online historical inquiry tasks: A case study of two secondary school classrooms. *Computers & Education, 50*(4), 1394–1410

Loh, B., Radinsky, J., Gomez, L., Reiser, B., Edelson, D., Loh, B., et al. (2001). Developing reflective inquiry practices: A case study of software, the teacher, and students. *Designing for science: Implications from everyday, classroom, and professional settings* (pp. 279–323).

Mayer, R. E. (2003). *Learning and instruction*. Prentice Hall.

Peterson, M., Morrison, D., Cram, K., & Misanchuk, E. (1996). CMC: An agent for active learning. In *Proceedings of 12th Annual Conference on Distance Teaching and Learning: Designing for Active Learning* (pp. 7–9).

Razikin, K., Goh, D. H., Theng, Y. L., Nguyen, Q. M., Kim, T. N., Lim, E. P., et al. (2009). Sharing mobile multimedia annotations to support inquiry-based learning using MobiTop. *Active media technology: 5th International Conference* (pp. 22–24). Beijing: Springer.

Roberts, M. (2003). *Learning through enquiry: Making sense of geography in the key stage 3 classroom*. Sheffield: Geographical Association.

Roberts, M. (2010). *Geographical enquiry*. The Open University, http://www.open.edu/openlearn/education/teaching-secondarygeography/content-section-0
Vavoula, G. N., & Sharples, M. (2003). Putting order to episodic and semantic learning memories: the case for KleOS. In *Proceedings of HCI International 2003* (pp. 22–27).
Wentzel, P., van Lammeren, R., Molendijk, M., de Bruin, S. & Wagtendonk, A. (2005). *Using mobile technology to enhance students' educational experiences*. Educause Center for Applied Research (ECAR) Case Study 2, 2005, Boulder, Colorado.
Yeh, R., Liao, C., Klemmer, S., Guimbretière, F., Lee, B., Kakaradov, B., et al. (2006). ButterflyNet: a mobile capture and access system for field biology research. In *Proceedings of CHI'06* (pp. 571–580). New York, NY: ACM Press.
Zurita, G., & Nussbaum, M. (2004). Computer supported collaborative learning using wirelessly interconnected handheld computers. *Computers and Education, 42*, 289–314.

Faisal Aman was a former humanities educator at a secondary school in Singapore and currently works at the National Institute of Education (NIE) as a Research Associate, looking at professional development programmes and higher degrees. His previous papers focused on enhancing the understanding of geography in secondary schools by looking at cross-disciplinary approaches. His research interests include geographies of professional work and teacher assessment, andragogy and geographical assessment.

Boon Kiat Tay is an experienced educator at Marsiling Secondary School, Singapore. He has been a teacher for seven years and has special interests in the use of technology for teaching geography in the field.

Chapter 10
High-Speed Mobile Telecommunication Technology in the Geography Classroom

Shahul Hameed, Pillai Vidhu and Tan Xue Ling Sherlyn Theresia

Abstract Digital devices such as handphones and computers allow teachers and students to access the functionality beyond telecommunication. For instance, students and teachers are able to interview experts through videoconferencing. With these advancements in technology, teachers need to be innovative and creative to engage students, interest them and allow them to see the relevance of their studies to the real world. As such, the authors have introduced the use of high-speed Internet connectivity mobile phones for the learning of the humanities subject, such as geography. This project started in 2010 and received positive feedbacks through surveys and examination results of students who have experienced mobile technology. The use of mobile phones enables students to conduct live interviews and walkthroughs of geographical sites, bringing the world to the classroom as the phone is connected to the classroom projector. Students also engage in cooperative teamwork, analytical thinking and Socratic questioning as they deliberate on the topic and craft questions in groups for the interview. During a videoconferencing session using mobile phones, 49 Secondary Three and Four students interviewed a NIE professor on the flora in Bukit Timah Nature Reserve using questions they had crafted in groups as they studied the characteristics of natural vegetation in their syllabus. Another example was an interview with the owner of a crocodile farm by Secondary Two students on how high-tech farming takes place in Singapore. The portability of the mobile phones not only allowed the interviewee to answer the questions asked but also gave students the opportunity to see how the crocodile farm looked like without leaving the classroom. From another perspective, students could also go out of the classroom, interviewing people and sharing the process and knowledge with their peers live time through a mobile phone. The use of

---

S. Hameed (✉) · P. Vidhu (✉) · T. X. L. Sherlyn Theresia (✉)
Pierce Secondary School, Singapore, Singapore
e-mail: shahul_hameed_kuthubudeen@moe.edu.sg

P. Vidhu
e-mail: pillai_vidhu@moe.edu.sg

T. X. L. Sherlyn Theresia
e-mail: tan_xue_ling_sherlyn@moe.edu.sg

high-speed mobile devices allows the authentic learning of topics in a humanities subject such as geography.

Introduction

We now live in a digital world where students and teachers are connected through digital devices such as mobile phones, tablets and computers. *With the advent of fourth-generation (4G) mobile phones providing high-speed Internet access, conventional* methods of teaching are challenged by the appeal the former has on our students. This encourages teachers to rethink the current pedagogues and to be innovative and creative in different ways to engage our students so that they can have meaningful learning that empowers them to engage issues in the real world. Studies have shown that students have used mobile devices to learn geography well (Sharples et al. 2005; Nguyen et al. 2008). Studies involving mobile devices and smartphones in the field setting by amateur naturalists have also been previously recorded. As such, the authors have introduced the usage of mobile phones for the learning of geography, in the geography department at their school.

Challenges in Teaching

From a pupil's point of view, geography, like other humanities subjects, has been a subject associated with "large amounts of content". In a sense, there is a need to move students from acquiring knowledge to the development of skills required to acquire the knowledge (Favier and van der Schee 2009). Due to the nature of these subjects, it has been a challenge for students to remember the material and at the same time apply what they have learnt in their examinations. This in turn leads to the pupils achieving their potential during examinations. There is also an observed decline in interest in humanities subject in the school that the authors are teaching in.

With this challenge in mind, teachers attempt to make learning more engaging using a plethora of traditional methods. While some teachers use the interactive and presentation elements of powerpoint slide shows, videos and cooperative group work, other teachers organize learning journeys by themselves or through services provided by vendors, both locally and overseas. This allows students to experience and visit the sites that they see and learn in their textbooks. However, learning journeys need to be planned in advance and many factors need to be considered, such as security and availability of budget, which may in turn limit the scope and the size of the group.

This chapter presents the efforts to revolutionize the way the humanities subjects are taught by the authors in their school. The authors suggest to enhance the current methods of student engagement, so students can bring the world into the classroom by using the latest wireless mobile telecommunication technology (Goodchild 2007).

The authors documented the use of mobile technology in their humanities classes with the initial goal of researching ways to improve students' grades in the humanities subject examinations. The research spanned months and involved frequent focus group discussions among members of the humanities department. This chapter presents the findings on using technology in the geography classroom, among related examples from the social studies and history classrooms.

At the time of writing the chapter, the authors were reporting on using the third-generation (3G) wireless mobile technology. While the authors are mindful that technology has advanced and the wireless speeds have far exceeded 3G mobile technology, the chapter will be exploring the affordances of high-speed mobile access to the Internet rather than whether there is a difference in speed of access between a third or fourth-generation device. The key affordances of using high-speed devices were that the authors were able to conduct live interviews and walkthroughs of historical and geographical sites through video calls which were beamed back into the classroom.

An example of a high-speed mobile phone enabled lesson begins with the teacher making a video call to an interviewee who may be in another location in Singapore. The teacher's mobile phone would be connected to the classroom projector so that students would be able to see and hear the interviewee during the call. During the interview, students will be invited to ask questions regarding the topic on hand. Responses given by the interviewee would then be taken down by pupils in their worksheets. The interviewee also has an option to showcase his cultural/geographical site, and he/she is working on the video call. The experience for the pupils engaged in a mobile-enabled lesson is both exciting and enriching (Chang et al. 2012).

What the authors are essentially doing was videoconferencing. The technology to videoconference for education has been around for at least a decade (Hurley et al. 1999). All that is required are for two computers with webcams attached and the necessary software to conduct a live video stream. However, this set-up might not be suitable all the time due to the nature of the people that the authors intended to interview. For example, the Singaporean that we interviewed who lived through World War 2 is almost 80 years old. Their knowledge and competency in using a personal computer, attempting to navigate around the software and adjusting the webcam might be cumbersome and tedious for the interviewee. In addition, the person may not have the necessary equipment to conduct the videoconference which means that the teachers will not only have to provide all the equipment but at the same time train the person in using teleconferencing software such as Google Hangouts or Skype.

However, mobile phones provide a technically easier alternative for videoconferencing. All the interviewee has to do is just answer the video call initiated by the teacher at a predetermined time and the interview can proceed. In addition, the portability of phones allows interviewees to "walk around" their workplace to showcase their artefacts or products on display. For example, when we interviewed a crocodile farmer in a lesson on high-tech farming in Singapore, he was able to show us the pens that housed the crocodiles. Similarly for the interview conducted with a historian working at the National Archives, she was able to showcase some artefacts on display at the location. This would not have been possible if a computer was used for the videoconferencing. Using mobile phones for this project was far more beneficial and enriching for the pupils.

Potential of Mobile Technology in the Classroom

Mobile phones offer great potential advantages as a tool for teaching and learning. From one perspective, it would allow the classroom teacher to connect the pupils with experts from different walks of life. Live streaming could allow students to interview a guest speaker without the hassle of either party having to travel to meet. Through this experience, the teachers are also able to stretch the capacity and creativity of our pupils by asking them to explore various ways in interacting with the interviewer on the other side of the phone through Socratic questioning and systematic analytical thinking processes.

In particular, mobile technology supports pedagogies for the humanities subjects of geography, history and social studies through enhancing the engagement and interactivity for the students. With mobile technology, students are able to relate what they learn beyond the textbook and classroom with the environment they live in. This makes the humanities subjects come alive and become more relevant for students.

In particular, a high level of **cooperative teamwork** among students was necessary to fulfil the performance tasks when using mobile technology. The various teams of students would deliberate on selected topics, identify people to interview, plan the timing of the live interviews, craft questions for the interview session, conduct the interview, sieve out pertinent information from the interview and make an oral and written presentation for assessment.

Students' creativity was harnessed by asking the pupils to interview, in some instances, their own parents and relatives for topics that were close to their heart. For example, the topic on "Patriotism in Singapore" involved almost all the parents of the classes to be interviewed in a span of a one-hour period for the lesson.

From another perspective, the teacher would also be able to assign groups of students using their mobile devices to record their out-of-school learning journeys or interviews of external resource persons to bring back to share with their classmates. This provided a rich source of contextualized information for students to work on.

Conceptualization and Development

As part of the longer-term action research on how to increase student engagement and improve student performance, the authors had initially planned a lesson using mobile technology which involved cooperative learning strategies. The lesson was followed by a student survey their perceived level of engagement in the lesson, and their self-perceived interest generated on the topic based on the lesson as well as some general feedback. Many students cited the use of videos and media as a source of engagement for them. The authors wanted to extend this and developed a plan to use mobile technology and cooperative learning in the classroom. This project started in 2010 and its benefits have since been seen through surveys conducted on students and the examination results of students who have experienced mobile technology. The use of mobile phones enables students to conduct live interviews and walkthroughs of geographical sites, bringing the world to the classroom as the phone is connected to the classroom projector. Students also engage in cooperative teamwork, analytical thinking and Socratic questioning as they deliberate on the topic and craft questions in groups for the interview.

Implementation

The implementation process involved the purchase of the mobile phones and the mobile data plans from a local Internet service provider. We particularly selected a model that allowed the phone to be connected to the classroom projector. Once the hardware was finalized, we collaborated with National Heritage Board who sent representatives to our schools to train teachers and students on the interview techniques and etiquette. This was necessary as pupils may have to interview the elderly which requires delicate skills in questioning. Students and teachers were also trained in Socratic questioning techniques. From the trial-and-error process in our teachers' prototyping of this approach, they have developed structured lesson plans for the students. Pre-planned questions for inquiry have been crafted which provided focus for lessons. For online interview sessions, the pupils were able to interact with the interviewees effectively. Professional ideas and personal viewpoints were effectively shared and authentic learning took place in their own classroom at their own pace. Teachers would work in pairs to carry out the lessons. One teacher would focus on finding and dealing with interviewees while another teacher would focus on the hardware set-up and running of the lesson. In this way, teachers leverage on each other's strengths and worked in teams to provide an enriching lesson.

Using High-Speed Mobile Technology in the Classroom

The team has used high-speed mobile technology in teaching various topic areas. The team has also created a checklist that shows a new user how to carry out a high-speed mobile lesson (Annex A), and s*ome lesson plans that have been used by the team can be found in Annex B*. Some of the Humanities Project topics that have been carried out through the use of high-speed mobile technology include:

Japanese Occupation—An interview with a Singaporean who has been through WW2.

High Tech Farming—An experience of a crocodile farmer in Singapore.

Terrorism and its Impact on Singapore—An interview with an NUS Professor

Traffic Conditions in Singapore—An interview with an LTA representative

History of Singapore—An interview with a historian from National Heritage Board.

Bukit Timah Nature Reserve—An interview with an NIE Professor

A High-Speed Mobile Lesson—Bukit Timah Nature Reserve

The team's high-speed mobile enabled lesson reported in this chapter was conducted for a combined class of 40 Secondary Three Elective Geography students and 9 Secondary Four Pure Geography students on the topic of natural vegetation. Preparation for this lesson started more than a month before the actual lesson when an interview with Professor Kalyani Chattterjea who is an expert on the flora and fauna at Singapore's Bukit Timah Nature Reserve was set up. After the lesson, a survey was also conducted on the students after the lesson to gather their feedback.

Preparation

At that time, the Secondary Three Elective Geography students were learning the topic on natural vegetation, and it seemed the most appropriate topic for a high-speed mobile lesson as the Bukit Timah Nature Reserve could be used as an example of how trees in a tropical rainforest adapt to the natural environment. With this topic in mind, Professor Chatterjea who is known to be an expert in this field and was thus the best interviewee for this topic was contacted. A meeting with her was arranged where she was provided a mobile phone for the lesson and the team explained the lesson outline to her.

In the classrooms, students used cooperative learning strategies to craft questions they wanted to ask Professor Chatterjea. These questions were prepared by the students with the guidance of the teachers to enrich students' learning on the topic and were refined several times to ensure its clarity and suitability for the interview. The lesson plan for this lesson can be found in Annex C.

Implementation

Before the lesson, students were briefed on the expectations of the lesson and student representatives chosen to interview Professor Chatterjea during the lesson. As student representatives took turns to interview Professor Chatterjea, the rest of the class listened attentively to the answers provided and diligently wrote them down in the worksheets provided. As students learnt more about the flora at Bukit Timah Nature Reserve, they were reminded of the concepts learnt and they were introduced to new concepts not found in the textbook.

Post-lesson

After the lesson, a survey was conducted to gather students' feedback. Through the surveys, it was seen that students enjoyed the lesson and found it engaging and enriching. Many looked forward to more of such lessons in the future. A problem, however, was the difficulty in hearing the interviewee due to technical issues.

Qualitative Feedback and Quantitative Results

Since the use of high-speed mobile technology, the passing rate of all the humanities subjects has surpassed the national level results. The team believes this can be attributed to the strong interest in Humanities that was cultivated over the years. The cultivation is due to the variety of projects the team has implemented to improve interests in humanities subjects, and the use of high-speed mobile technology in the humanities classroom is one of them.

Not only has high-speed mobile technology expanded the learning space of the children but also it has made learning more authentic. The immediacy of the experience has made lessons come alive and allowed them to appreciate the relevance of Humanities subjects.

Pupils could interact with other professionals without boundaries. We have begun to create classrooms without walls, in which the pupils have the opportunity to interact with people from all walks of life. This sharing of other individuals' life experiences has added value and greater meaning to lessons. High-speed mobile technology also allowed for instantaneous feedback and interactivity. Pupils had the satisfaction of getting answers from people who have experienced the topics. This opportunity leaves a lasting impression on many young minds.

When we embarked on this journey, we faced difficulties when many of the interviewees had concerns about finding time to travel to school to meet the pupils. High-speed mobile technology has allowed speakers to be involved in all class discussions without being physically present and without taking time off from their work. If these interviews had been conducted without the use of technology, speakers would have had to address a large assembly in a huge hall—which is a far less conducive setting. On the other hand, if we had organized a mass dialogue session, it would not have given the opportunity for all the pupils to interact with the interviewer online.

High level of pupil engagement was evident in many classes through the variety of questions asked by the many pupils. Even though there were structured inquiry questions prepared by teachers, we observed that many pupils took the opportunity to raise impromptu questions in response to points raised during the live interviews.

After the lessons, pupils were seen engaged in discussions out of the classroom about the topics that were discussed via the high-speed mobile lessons. The points below summarize the learning of the two groups involved.

Pupils' Learning Outcomes

1. Better understanding of topics in History and Geography
   Pupils are now able to understand the topics such as Japanese Occupation, Independence of Singapore, Terrorism and its implications on Singapore as well as Geography topics such as High-Tech Farming in Singapore.
2. Higher interest level in Humanities subjects.
   This is evident in the higher number of pupils taking Pure Humanities subjects and the value-added results achieved from the examination results for the Combined Humanities subject.
3. Higher sense of ownership in learning.
   Pupils are also seen to be discussing the topics beyond classroom, when planning for the live conferencing sessions.
4. Interviewing Etiquette.
   Pupils are now better able to communicate effectively with professionals knowing phone and conversational etiquette.
5. Analytical Thinking and Socratic Questioning
   Pupils are now better able to analyse questions that they have thought for themselves as well as questions posed by the teacher. The thought processes are rigorously kindled when they further discuss issues brought up during the high-speed mobile oral interviews.
6. Team Learning
   Team learning is encouraged as pupils have to work in groups to process the questions to be asked after a lesson and eventually come together to see whether the questions have been answered. The process comes to a celebratory close, where pupils then present their data to the rest of their peers in a way that summarizes key learning points of the lesson after the high-speed mobile interviews.

Teachers' Learning Points

1. Creative ways of engaging pupils to better understand a topic
   Using high-speed mobile technology has challenged teachers to go out of the way to plan lessons in a new way, e.g. inviting resource persons to the cyber classroom. This has created a fresh opportunity for engaging the pupils in an authentic learning context.
2. Team Teaching
   Teachers are able to collaborate on designing lessons. For instance, this allowed more than one class of pupils to engage the invited interviewee through the high-speed mobile conferencing.
3. Better understanding of pedagogical tools for effective teaching
   Teachers have the opportunity to apply effective teaching strategies utilizing the Mobile technology that engage the minds and hearts of the people for meaningful teaching.

Discussion

The use of high-speed mobile technology in the classroom is a long-term plan that has been fully adopted by the team. After the initial trial, there were plans for each humanities teacher to conduct at least two high-speed mobile lessons per semester. New teachers who have been posted into the department were also briefed on the use of high-speed mobile and are trained to adopt it.

The potential for this technology to be used in other areas of education is promising. As an extension, the team plans to develop a high-speed Source-Based Question Trail, which comprises of an outdoor trail and learning activities. Students would be tasked to take photographs of various locations and use them as sources to complete the questions. In addition, students would be able to conduct interviews and record the session for use in group projects or presentations in class. The Mathematics and English Departments have also shown keen interest in high-speed mobile technology and are exploring ways to incorporate this into their curriculum.

As technology improves, so would the capabilities and quality of the videoconferencing and this would open up more channels for the team to explore the use of this technology within the classroom, providing ample opportunity for the long-term sustainability of this project.

This project can be scaled up with little or no refinements. We initially tested it out on Secondary 2 students with the Japanese Occupation lesson. Following that we also used high-speed mobile lessons on Secondary 1,3,4 and 5 students. The team did not change the content, but only in the way they taught the content. The team has also used **Skype** to bring historical sites such as the 44th Civilian War Memorial Service from Beach Road to the classroom.

Conclusion

Using high-speed mobile technology is largely evident in our lives today, with the potential for creativity, and applications for numerous purposes such as teaching and learning (Chang et al. 2012). As teachers in this new generation, it is essential for us to keep thinking of new ways to engage and enrich our students. In fact, teachers should communicate and collaborate to explore how we can use, "technology … [as it] allows teachers to network and share each other's practices" (Parkinson 2013, p. 193). Using high-speed mobile technology is something very familiar to most of them, and we should leverage on it to enhance our teaching and our students' learning. Even though there are obstacles such as the demands on time, effort and resources in using a tool such a videoconferencing (Hurley et al. 1999), the benefits from using high-speed mobile Internet would outweigh the constraints. Indeed, students learn better in authentic learning settings where they have to solve real-world problems (Barron and Darling-Hammond 2008). We also took into consideration that learning geography is about understanding the environment and humans, their interrelationships and interaction at different scales. Given technological advancements, it is unavoidable that learning should not be restricted by physical space and that technology can bring the world to the classroom. With this in mind, the teachers on the team will continue to use high-speed mobile technology in the classrooms to enhance students' interest in the subjects.

Annex A

Checklist for conducting a Mobile lesson in your class

Ten things to look out for:	Tick (√)
You must have at least two mobile phones with TV output function that can be connected through the VGA panel for projection to pupils	
You should be able to have a mobile data reception in the classroom that the interview is planned for	
Identify the lesson that you want to conduct It must be a lesson that can allow interactions between an interviewer and the pupils It must be a lesson where pupils can learn from an interview, more on the particular topic that has been chosen Craft out questions that will be used for the interview	
Identify an interviewee, who can contribute to the development of the topic chosen	
Liaise with interviewee for the time slot that the lesson is to be conducted	
Arrange with him/her to pass one set of the mobile phone. Also pass him/her the set of interview questions and get him/her ready	
Do a test run of the mobile data reception, before the actual day of the lesson, by calling interviewee from the classroom to check for reception	

(continued)

(continued)

Ten things to look out for:	Tick (√)
Prepare pupils who will be the interviewers	
On the actual day of the lesson, connect your mobile phone set to VGA panel. Test its readiness Carry out the lesson as per planned	
After the lesson, collect phones from interviewee Out of good gesture, bring along a small gift	

Annex B

Participant's Lesson Plan on 3G Technology in the Humanities Classroom

Topic	Lesson objectives	Scaffolding	Who to interview	Interview questions	Follow-up
Social Studies: Conflict between countries (Case study of Pedra Branca)	. list 3 causes of conflict . state how Singapore has defended Pedra Branca . state the value of Pedra Branca	. pre-interview knowledge on the conflict . videos on cause of conflict . newspaper articles. Map. Info about ICJ	Officers from Ministry of Foreign Affairs or Lecturers from NUS/ Institute of SEA Studies	1. what is the value of Pedra Branca to Singapore? 2. Will this affect Singapore's relationship with Malaysia 3. How did/ have we defended Pedra Branca?	. Quiz on Pedra Branca
History: Life in Singapore during the Japanese Occupation	Students will be able to: . gather authentic/ first hand info on life during the Japanese Occupation . compare different viewpoints of the	. video documentary (Rape of Nanking) . newspaper, photographs	. bringing in senior citizens who have lived through the Occupation (3 different races) . Professor who is an expert on Japanese Occupation (3G)	1. was everyone equally treated during the Japanese occupation? 2. why was there discrimination? 3. why were they so harsh? 4. was life better off under the British or the Japanese?	Worksheets, tests, mind maps

(continued)

(continued)

Topic	Lesson objectives	Scaffolding	Who to interview	Interview questions	Follow-up
	Japanese Occupation				
Geography: Geography of food	Pupils will be able to: . explain the changing food preferences of consumers in Singapore . explain the impact of changing food preferences on the types of products being sold in Singapore	. ask students about their favourite foods. Show pictures of different kinds of foods . compare food preferences then and now	Personnel from AVA or Nutritionist such as Devagi Shanmugam, who runs her own Indian Fusion Café at Race Course Road, who is also the host of a programme at the Asian Food Channel	1. what type of foods Singapore imported in the 60s 2. what type of foods Singapore imports now? (the difference) 3. what are some of the countries that we import our foods? Which countries do we import most?	Consolidate and summarize with students. Homework. Reinforce lesson objectives

Annex C

Lesson Plan on 3G Lesson—Bukit Timah Nature Reserve					
1	Teacher's name:		Date:	Time:	Class:
	Mrs Chung, Miss Sherlyn Tan		10 April 2012 (T2W4)	1215–1310	3E1/4E1
2	SUBJECT: Geography		TOPIC: natural vegetation		VENUE: Cave 2
3	SPECIFIC LEARNING OBJECTIVES: Pupils will be able to…	1	Explain 4 ways how plants in the tropical rainforest adapt to their environment		
		2			
		3			
4	LESSON:	Time/ duration	TEACHING— LEARNING ACTIVITY		Resources/key concepts:
	PRE-LESSON	1 month	About one month before the lesson, students would have been asked to think of questions (in		Students' questions. This pre-lesson prepares students for

(continued)

(continued)

Lesson Plan on 3G Lesson—Bukit Timah Nature Reserve				
			groups of four) relating to the characteristics, adaptations and attractions of the tropical rainforest and submit them to the teacher, who will collate them. These questions would be given to Dr Chatt from NIE who is an expert on Tropical Rainforest, especially the Bukit Timah Nature Reserve, for her reference and preparation for the 3G lesson that is going to take place in this lesson. Dr Chatt will be asked to provide any additional information on the tropical rainforest to the students so as to heighten their knowledge on the tropical rainforest in Singapore, namely the Bukit Timah Nature Reserve	the upcoming 3G lesson
4.1	START:	10 min	Students will be given a field lesson in the area outside the school where they will be brought out of their classrooms to see and feel the different characteristics of trees in a tropical rainforest. This experience will enhance their understanding on the adaptations of plants	Conducted by Mrs Chung
4.2	MIDDLE:	35 min	The teacher prepares students for the 3G conference with Dr Chatt. Selected students from each group have been tasked to pose a question to Dr Chatt during the conference. These students will take turns to pose Dr Chatt	3G phones, worksheet, question cards for each group, worksheet for every student

(continued)

(continued)

Lesson Plan on 3G Lesson—Bukit Timah Nature Reserve				
			their assigned question. At the same time, each student will be given a worksheet which they are expected to complete using the answers given by Dr Chatt during the 3G conversation	
4.3	END	10 min	After the lesson, teacher will get different groups of students to present the answers given by Dr Chatt verbally as a summary. Students will then be tasked to do a concept map and an essay (individual assignment) on the adaptations of plants in the tropical rainforest based on the lesson and answers provided. Students will have to submit their individual assignment the next lesson	Essay assignment Questions

References

Barron, B., & Darling-Hammond, L. (2008). *Teaching for meaningful learning: A review of research on inquiry-based and cooperative learning*. Book Excerpt. George Lucas Educational Foundation.

Chang, C. H., Chatterjea, K., Goh, D. H., Theng, Y. L., Lim, E., Sun, A., et al. (2012). Lessons from learner experiences in a field-based inquiry in geography using mobile devices. *International Research in Geographical and Environmental Education, 21*(1), 41–58.

Favier, T., & Van Der Schee, J. (2009). Learning geography by combining fieldwork with GIS. *International Research in Geographical and Environmental Education, 18*(4), 261–274.

Goodchild, M. F. (2007). Citizens as sensors: The world of volunteered geography. *GeoJournal, 69*(4), 211–221.

Hurley, J. M., Proctor, J. D., & Ford, R. E. (1999). Collaborative inquiry at a distance: Using the Internet in geography education. *Journal of Geography, 98*(3), 128–140.

Nguyen, Q. M., Kim, T. N., Goh, D. H. L., Theng, Y. L., Lim, E. P., Sun, A., et al. (2008, September). TagNSearch: Searching and navigating geo-referenced collections of photographs. In *Proceedings of the 12th European Conference on Research and Advanced Technology for Digital Libraries* (pp. 62–73). Springer.

Parkinson, A. (2013). *How has technology impacted on the teaching of geograhy and geography teachers? Debates in geography education* (p. 320). Abingdon: Routledge.

Sharples, M., Taylor, J., & Vavoula, G. (2005, October). Towards a theory of mobile learning. In *Proceedings of mLearn* (Vol. 1, No. 1, pp. 1–9).

Shahul Hameed A Geography and History teacher by training, Shahul was the Head of Department (Discipline) at Braddell-Westlake Secondary School, Singapore, in 2005, and from 2006, he held concurrent posts as Head of Department (Humanities) and Year Head (Upper Secondary) at Peirce Secondary School. During this time, he was also appointed Chairman of the school's Curriculum Development Committee, a committee set up to spearhead the school's drive towards providing a holistic learning experience for every child in school.

Pillai Vidhu A History and English teacher by training, Vidhu was the Subject Head (Innovation) and Year Head (Upper Secondary) at Peirce Secondary School. His job was primarily to spearhead and encourage innovative programmes in the school.

Tan Xue Ling Sherlyn Theresia A Geography and Mathematics teacher by training, Sherlyn was the Upper Secondary Coordinator (Geography) at Peirce Secondary School. During this time, she was also an ICT mentor and spearheaded the Humanities Week and Earth Day at Peirce (2012) which got the school into the Singapore Book of Records.

Part IV
Reflecting on Factors that Support Teaching and Learning Geography Beyond the Traditional Classroom

Summary of Chaps. 11–13.

As the authors reflected on the purpose of extending teaching and learning geography beyond the classroom, there was a consensus that geography must be lived and experienced. It has to be relevant to the lives of our students and this cannot be done solely through textbooks and pen and paper tasks. The key question then is to examine the possibility of how field inquiry and the use of ICT can engage students in critical and reflective thinking. Beyond integrating both fieldwork and ICT (Chap. 11), the school environment is an important factor in supporting learning geography beyond the traditional classroom (Chap. 12). However, fieldwork and ICT take us beyond learning about geography as they also help us develop a person holistically.

Chapter 11 examines the use of an inquiry approach in both fieldwork and classroom teaching through a social media tool. The aim is to engage students in "independent learning, critical thinking, reflective thinking and inquiry." (Chang and Seow 2010). This chapter narrates the experience of the geography educators in a Secondary School in carrying out a field inquiry on the topic of river channel management. In addition, the social media tool of a blog was used concurrently to prepare students for the pre and post field trip phases. Chapter 12 discusses the role school leaders play in enhancing teachers' comfort level in using ICT for teaching and learning. While there has been significant understanding on the types of school leaders, the roles they play and the areas of change they make to enhance technology integration in schools, the authors found that principals and heads of department (of instructional programmes) play a major role in building "a strong culture of sharing within departments to provide support for teachers to equip them with the relevant ICT integration knowledge and skills". This chapter will also discuss the different type of ICT-enabled lessons commonly used by teachers.

Chapter 13 is a reflective summary of all the chapters. It begins with a reiteration on why fieldwork is important for geography and the importance of subject disciplinary thinking in geography fieldwork. This is followed by the importance of faculty members bringing their own research to the classroom and having students participate in the geographical process of constructing a large raingarden in NIE campus. The students' experiences are captured vividly in their personal "daily log" during their fieldwork experience. It then highlights that there is no escaping from the use of ICT in our children's learning experiences, as it is an integrative part of our students' reality. Therefore it is important to examine how ICT can be used meaningfully, in matching the learning activities we design for student's needs and the geographical knowledge and skills that have to be taught.

The chapter closes nicely with a reminder that Geography is about understanding the interaction between nature and human, their interrelationships and interactions across scales, and that learning should take place beyond the classroom and into the field, and even to venture into "places" beyond.

Reference

Chang, C. H., & Seow, T. (2010). *Field inquiry for Singapore geography teachers*. Paper presented for SEAGA 2010, Hanoi November 23–26 2010. Online Proceedings.

Chapter 11
Social Media—A Space to Learn

Frances Ess

Abstract The Ministry of Education's (MOE) third master plan for information and communication technology (ICT) in Education (2009–2014) has the vision to enrich and transform the learning environments of our students. The plan aims to allow students to develop competencies for self-directed and collaborative learning through the effective use of ICT as well as to become discerning and responsible ICT users (Ministry of Education, Singapore, 2013). As geography educators in schools, we have limited classroom interaction with our students. We have to conduct tests and sacrifice lesson time for camps, public holidays and school events. It is a challenge for geography educators to conduct inquiries in class, as he or she has to balance between covering the syllabus and engaging students in discussions. One way to overcome the challenge of time is through the use of ICT to engage students beyond the classroom space. By using social media like Facebook and Wordpress, geography educators and students now have an opportunity to continue their discussions beyond the classroom and develop competencies for self-directed learning. In addition to the use of ICT, there is a possibility of redefining the concept of a classroom as it can now be seen not only as a physical space limited by structural walls. Thus, ICT affordance allows geography educators to radically expand classroom space and time through immersive experience brought into the classroom through videos and extending learning beyond scheduled time. Moreover, the use of ICT has also redefined the concept of teaching as information dissemination to the cultivation of skills, values and attitudes for a lifelong process of learning. This chapter aims to explore how one geographer educator uses social media to engage her students and identify challenges that the use of ICT might bring. She has conducted a field inquiry on River Studies with students using Wordpress as a platform to share background reading, constructing knowledge and reflections (see http://seriesess4.wordpress.com/). She will share with readers the problems, pitfalls and possibilities that geographers using this method will encounter. An analysis of the reflections posted on the site by students will be used first to see if attitude and mindset point towards self-directed learning. A pretest and

---

F. Ess (✉)
Mayflower Secondary School, Singapore, Singapore
e-mail: frances_ess@moe.edu.sg

a posttest were conducted to evaluate student performance so as to examine the effectiveness of using ICT to engender deep learning.

Introduction

Ever since the introduction of the first master plan for information and communication technology (ICT) in Education in Singapore in 1997, informational technology has made it possible for educators to work and connect with students 24/7. The learning space has expanded beyond the classroom. ICT has made information readily available and ubiquitous. This could increase our students' desire to learn, educators could link and guide students to various information sources and promote self-directed learning. Therefore, how educators integrate ICT into teaching and learning has become a point of interest for many educators.

In 2013, a revised Geography syllabus was implemented for the GCE "O" levels in Singapore where the emphasis is on the use of an enquiry approach in both fieldwork and classroom teaching. This approach is expected to engage students in "independent learning, critical thinking, reflective thinking and inquiry". (Chang and Seow 2010). As part of the preparation for the implementation of the revised Geography syllabus, the geography educators in Mayflower Secondary School decided to carry out a field inquiry on the topic of river channel management. In addition, the social media tool of a blog was used concurrently to prepare students for the pre- and post-field trip phases. This study aims to examine the possibility of marrying these two processes (field inquiry and the use of an ICT tool) so that geography educators can be more effective in engaging students in critical and reflective thinking.

Purpose and Relevance of Study

The Bishan-Ang Mo Kio (BAMK) Park was opened on 17 March 2012. One of the star attractions is a meandering "Kallang River" running through the park, which has been carved out of a concrete canal, flows into Marina Reservoir. The river is expected to slow the flow of rainwater and thus help to prevent the flood in the downtown area. This provided an excellent opportunity to conduct a field inquiry on river channel management for the following reasons:

Proximity to the school. The BAMK Park is only a five minutes' -walk from the school, and thus, a field inquiry lesson could even be conducted during double period lessons.

The BAMK Park offers students an opportunity to study a river channel management that is different from what has been expounded in textbooks. In most textbooks, river sectioning and river channelization are proposed as a form of flood control.

However, a linear, utilitarian, concrete drainage channel has been redeveloped into a natural meandering river at the BAMK Park. Students are provided with an opportunity to understand that textbooks do not necessarily represent reality.

Literature concerning the redevelopment of the park can be found in cyberspace, and we wanted students to learn to judge whether these sources are useful in helping them to understand the issue of river channel management.

Many students walked through the park to go home in the Sin Ming Drive estate, and this field inquiry would spark their curiosity about the rationale for the redevelopment of the channel into a river.

Field Inquiry at BAMK Park

Structure of the Field Inquiry

This field inquiry was carried out from July 2012 to August 2012. Three sessions were conducted, the first for a group of Sec 5 Normal Academic students, the second for a group of Sec 4 Express Stream students and the last for peer leaders who were expected to conduct a similar field inquiry for their peers.

Phases of the Study

The field inquiry was divided into three parts (pre-field trip, field trip and post-field trip) using MOE's Humanities Inquiry Process as a guide (Ministry of Education, Singapore, 2016). Students will undergo the four stages of inquiry process, of sparking curiosity, gathering data, exercising reasoning and reflecting thinking (Roberts, 2003).

Pre-field Trip Phase

A blog entitled "You cannot step twice into the same river" (http://seriesess4.wordpress.com/) was developed before the students embarked on the pre-field trip phase.

The advantages of using social media to compliment the field inquiry are as follows:

Social media enables the sharing of information. It allows educators to upload relevant reading materials from various sources (text, photos, audio and video) for students to read as background information.

Students are provided a platform to study the reading materials in their own time. There is no wastage of paper as the reading materials are online.

Social media is dynamic as it allows students to post their replies and comments on the blog so that other students can read their comments and learn from them.

Instructions were placed in the blog to guide the students on the various process of field inquiry.
It also allows the educator to evaluate the students' level of understanding.

Social media is not subjected to a fixed schedule. Educators are no longer constrained by the school timetable (e.g. number of classes/periods per week). Both educators and students can now expand and enlarge the space to learn beyond the classroom.

A briefing was conducted in class to guide the students through the various processes of the field inquiry. In line with the revised Geography syllabus, students were first introduced to the overarching geographical inquiry question: "River Channel Management: Is it worth it?" They were then shown the page in the blog entitled the Big, Big Picture which helped them to frame the rationale for the field inquiry. (http://seriesess4.wordpress.com/big-big-picture/). Students were given one week to be engaged independently and prepare for the next two pre-field trip phases/sessions of Sparking Curiosity and Gathering Data.

Sparking Curiosity

For this activity, students watched a video on how the river has been transformed from a concrete channel to a natural river over the span of two years. The aim of the video was to spark the students' curiosity as to why the government of Singapore would spend $76.7 million to redevelop the park and to reintroduce a naturalised river back into the park.

Gathering Data

Here, instead of providing a list of reading materials for students to read up, we invite students to evaluate the various sources that have been given to them.

Questions included

From the two sources provided, which one do you think is more reliable? Give two reasons for your answer.

How useful is this source in helping you understand bioengineering?

Compare these two sources. Which one do you think is more useful for students who are studying about strategies about river management? Give one reason for your answer.

Not everything they read online was useful and reliable, and it is always good to sharpen students' evaluation skills. Therefore, students were trained in critical thinking through the gathering of data and preparing for the field trip.

Field Trip Phase

During this phase, students were brought to the BAMK Park. They were instructed to make observations of the Kallang River for 10 min. They were then asked to gather together and share questions that they formed after making their observations. Questions included:

Why are there pink eggs on the rocks?

Why are there red markers located at various locations?

Why are both hard and soft engineering methods used for this river?

Next, students were asked to form hypotheses about river channel management with specific references to this river. They were given time to collect data (e.g. sketched maps, measurement of the wetted perimeter and photographs) to test their hypothesis. Due to the constraints of time, students were only given 30 min for this phase of the field inquiry.

Post-Field Trip Phase

Students were given another week to work independently on the next two phases of the field inquiry.

Exercising Reasoning

Several questions were posted to test students' understanding of concepts that they have encountered during the field trip phase. Unfortunately, because marks were not allocated for this section, there were no responses to these questions. However, this neither means that students did not visit the page nor that they did not try out the exercises posted. (http://seriesess4.wordpress.com/reflection-2/)

Reflection

The students were more responsive when asked to post their reflections online. This could be due to the fact that the page asking for reflections had been structured so that student's responses are expected. (http://seriesess4.wordpress.com/reflection/)

Formative Assessment

A short class test was conducted one week later to evaluate if students have mastered the concepts and content encountered during this field inquiry.

Impact on Learning

The author has been using ICTs and social media since 2008 because it can empower educators and students and thus contribute to learning and achievement. It is based on this belief that the author has carried out an action research on the impact of ICT on student achievement using this field inquiry approach.

A group of 35 students went through the field inquiry as discussed in the previous section. A class test on the topic river management was given to this group of students (the experimental group) and another group of 36 students (the control group), who did not go for the field inquiry.

The key inquiry for this study is then to find out if students who participated in the field inquiry on river channelization performed better then students who do not participate in the field inquiry. The two sets of data were analysed to see if the students scored differed before and after the inquiry. While the author is cognizant that given the small sample size and the difficulty in determining the normality in the sample, the two sets of data seem to show that students' performance has changed before and after the inquiry. This observation, however, does not provide any suggestions as to why or how these students could have performed better than those who did not go for the field inquiry. An analysis of the reflections posted by the students in the blog could indicate or suggest possibilities as to why they had performed better.

From Student A:
> I feel that the field inquiry was conducted to allow us to have a better understanding of the river system and its processes. The field inquiry was helpful in the sense that it allowed me to get up close to the river, feel it and learn first-hand rather than reading the information from a textbook. While most of the information can be found on the internet, it cannot replicate the sense of smell and touch; therefore, the inquiry enabled me to construct new knowledge such as the texture of the sand in the river.

From Student B:
> It is actually amazing to see the big change, from the canal to the river. I have to say that they did a brilliant job!! It truly was a nice sight to see different types of plants, colourful dragonflies and small fishes in the man-made river. I was very curious about which direction was the river flowing!!! And when I went down for a closer look I only got more confused! But in the end, I figured it out by touching the water and felt the flow of water against my hand. This field trip sparked quite a few questions and made me understand more about rivers, like how the vegetation and gabions help prevent flooding and increasing the water capacity. After some research, I found out that the river is 3 km long and the river's capacity is 40 percent more than when it was a canal.

From Student C
> From the trip, I actually experienced something that the textbook did not mention; the reasons for the use of either hard or soft engineering bank protection methods. I could actually see that using both hard and soft methods will give the best results, and also why different protection methods were used at different parts of the river. This is something that could not be found in the textbook, and I am glad that I had gone for the trip as I learnt something that I would not have possibly known if I had not gone.

The above three reflections seem to suggest that students perform better in the class test because of the following reasons;

From Student A's reflections, we can see that he has gained a deeper understanding of the river system and its processes and this has enabled him to construct new knowledge such as "the texture of the sand in the river". The basic belief of constructivism is that knowledge is actively constructed by learners rather than transmitted by the educator. Learners are active knowledge constructors rather than passive information receivers (Jonassen 1991).

From Student B's reflection, words like "curious", "I figure it out", "sparked quite a few questions" and "I found out" seem to indicate that the field inquiry has provided a platform for students to engage in critical and reflective thinking. This in turn indicates that self-directed learning is taking place, and this could have contributed to the better performance of the students in the test.

Student C's reflection demonstrated the advantage of reading or going beyond the textbook. There seems to be a sense of ownership of the knowledge that the student has constructed and gained.

Problems, Pitfalls and Possibilities

Wang (2008) observed that with the rapid development of emerging technologies, the integration of ICT has increasingly attracted the attention of educators. However, he warned that a simple combination of hardware and software will not ensure that integration will automatically follow. There is a greater need to plan thoughtfully before geography educators start integrating social media into the curriculum.

Several problems were encountered while carrying out this project to integrate the social media tool (e.g. blog) into field inquiry. Firstly, it was assumed that students do not have difficulty assessing the blog because of the ubiquitous proliferation of the iPhone, iPad and iPod and that every student has access to a computer anytime and anywhere. This assumption does not hold true for all students, as some who are under financial assistance do not have the necessary resources to buy or subscribe to a data plan, let alone a computer. They would thus be required to make an extra effort to go to the public or school library where Internet access is free.

Secondly, it was assumed that students would have read the required materials that were posted online. Because students were trusted to read the materials before they embarked on the field inquiry, there was no way that I could check on their understanding or if they have read the necessary article. Of course they could post their problems and concerns online for the educator to answer but in this case, it was not forthcoming.

Thirdly, it was assumed that students would post their comments and responses on the blog just like they would hand in their homework on time. This requires a change of mindset from the students. Many students do not see social media (e.g. Facebook, Twitter and blog) as serious platforms where they can learn and collaborate with each other and educators to learn, to engage in critical thinking and to construct new knowledge. Thus, it was observed that students display a "touch and go" phenomenon where they post the required reflection and do not return to the site to see the comments that were put up by the educators. In addition, students tend to display a one-way street communication between educators and students. They seldom critically evaluate their peer's postings and reflections. Therefore, the geography educators need to establish protocols and a culture of behaviour through which to engage the students online. Hopefully, this would help to facilitate communications and discussions in the blog as it is a powerful space for students to learn.

Conclusion

Social media are transforming our lives at an exponential speed. Customer feedback is now obtained through blogs, advertisers use social media to build their brand and our students are immersed in social media constantly. Although I have encountered several problems that require remediation, I am inspired to continue using social media as a learning space for a couple of reasons. First, social media offers me more time to engage my students and cover more in-depth issues that I cannot cover in a conventional classroom. I hope to convince all the students to use social media as a complementary tool to learn. Secondly, it allows me to reach out to students with different learning styles, as I can upload videos, pictures and texts onto the blogs, so as to offer them a variety of resources for their own study at home.

Geography educators can no longer depend on the trusty chalk and board method alone to engage students. We should embrace social media as it offers an enlarged learning space for our students to engage with us and to challenge us with new ideas and concepts. If we do not enter this new learning space, we as educators are only limiting ourselves, and soon our students would find us obsolete and as interesting as the dodo bird.

References

Chang, C. H., & Seow, T. (2010). Field inquiry for Singapore geography teachers. Paper presented for *SEAGA 2010, Hanoi 23–26 Nov 2010. Online Proceedings*.

Jonassen, D. (1991). Objectivism versus constructivism: Do we need a new philosophical paradigm? *Educational Technology Research and Development, 39*(3), 5–14.

Ministry of Education, Singapore. (2013). *Development of 21st century competencies in Singapore*. Retrieved July 3, 2016, from https://www.oecd.org/edu/ceri/02%20Wei%20Li%20Liew_Singapore.pdf.

Ministry of Education, Singapore. (2016). *Social studies syllabus, upper secondary*. Retrieved July 1, 2016, from https://www.moe.gov.sg/docs/default-source/document/education/syllabuses/humanities/files/2016-social-studies-(upper-secondary-express-normal-(academic)-syllabus.pdf.

Roberts, M. (2003). *Learning through enquiry: Making sense of geography in the key stage 3 classroom*. Sheffield: Geographical Assoiciation.

Wang, Q. (2008). A generic model for guiding the integration of ICT into teaching and learning. *Innovations in Education and Teaching International, 45*(4), 411–419.

Frances Ess is an experienced educator at Mayflower Secondary School, Singapore. She has been a teacher for 30 years and believes in developing critical thinking skills in her students. She has six children and while she allows her children to climb trees, she does not allow them to play computer games or watch television programmes.

Chapter 12
Teachers' Comfort Level and School Support in Using Information and Communications Technology (ICT) to Enhance Spatial Thinking

Zhang Hua'an Noah and Tricia Seow

Abstract Recent rapid advancement of technologies has increased the sophistication of classroom instruction and provided a broad spectrum of new possibilities to develop and enhance students' learning. Based on a review of previous studies conducted in Singapore, it has been found that school leadership has some effect on teachers' comfort level in using ICT for teaching and learning. This chapter discusses how school leaders and middle managers influence geography teachers' comfort level in using ICT to enhance spatial thinking in students. Through surveys and interviews with ten practising geography teachers, the authors conclude that the middle managers play a vital role in sculpting an appropriate system and culture essential for teachers to feel comfortable when using ICT in their lessons. However, the chapter will also discuss the idea that geography teachers are mostly comfortable delivering only ICT-enabled lessons that involve 'spatial observation', and that school leaders and middle managers play a role in shaping these tendencies.

Introduction

Background of the Study

The advent and proliferation of Information Communication Technology (ICT) have brought new opportunities for transformation across the global education landscape. Education in the digital age has mapped new terrains for prevailing modes of pedagogic practices in what Resnick (2002) calls a 'learning revolution'. The advancement of technologies in recent decades has not only increased the sophistication of classroom teaching but also opened up a continuum of possibilities

---

Z. H. Noah (✉) · T. Seow (✉)
National Institute of Education, Nanyang Technological University, Singapore, Singapore
e-mail: NIE151153@e.ntu.edu.sg

T. Seow
e-mail: tricia.seow@nie.edu.sg

© Springer Nature Singapore Pte Ltd. 2018
C.-H. Chang et al. (eds.), *Learning Geography Beyond the Traditional Classroom*,
https://doi.org/10.1007/978-981-10-8705-9_12

to develop and enhance students' learning. In particular, its fundamental role in facilitating, supporting and strengthening learning goals continues to be broadly acclaimed by educators in this New Media epoch. (Tearle 2003b)

ICT has revolutionized the way teachers and students approach learning in the classroom. In Singapore, the Ministry of Education (MOE) has been actively exploring ways to leverage ICT to drive educational change in schools. To do so, the ICT master plan—which undergirds the collective thought leadership and strategic direction of the ministry's ICT enterprises—is reviewed every 5 years, so that the learning ecosystem in schools continues to be strengthened with the use of ICT (MOE 2016).

Early research on ICT use in teaching and learning sought to document the varying experiences of teachers delivering ICT-enabled lessons in Singapore. These initial studies mostly encapsulated the experiences of teachers using ICT to engage students' learning and enhance higher-order thinking in their respective subject areas (e.g. Tay 2002; Hung et al. 2003; Tan 2005; Looi et al. 2004). In comparison, scholarship today has critical focus on some of the reasons behind teachers' resistance to ICT-enabled lessons (e.g. Chen et al. 2012; Toh 2013; Liang et al. 2013; Tay 2014). These recent studies highlight the fact that the provision of even the most technologically advanced infrastructure does not necessarily ensure that teachers will be fully comfortable delivering their lessons with the use of ICT.

Enhancing teachers' comfort level in using ICT for teaching and learning begins with school leadership (Divaharan 2007). In particular, responsibility is vested on principals and vice-principals to build 'a strong culture of sharing within departments to provide support for teachers to equip them with the relevant ICT integration knowledge and skills. Also important are the middle managers in the school, for instance Heads of Department (of instructional programmes) and the Head of Department in ICT whose role is paramount in 'guiding the staff and heads of department to explore the use of technology in education' (Tan 2010, p. 12).

Whereas current scholarship offers a universal understanding of how school leaders and middle managers in Singapore influence the comfort level of teachers using ICT, this chapter aims to contribute to this understanding as a piece of empirical research conducted on teachers specializing in lower secondary geography education. Drawing attention to the principle geographical concept of space underpinning the teaching and learning of geography, the chapter examines the ways that teachers engage spatial concepts through a typical ICT-enabled lower secondary geography lesson and investigate whether teachers are comfortable delivering these lessons in ICT. Subsequently, questions relating to the types of support provided by school leaders and middle managers to help geography teachers feel comfortable using ICT during lessons and whether such leadership is a strong variable influencing their comfort level will be addressed.

Objective of the Empirical Study

Typically, empirical discussions focusing on the interfaces between school leadership, middle management, and teachers' comfort level in using ICT are often not situated in any particular subject. Consequently, subject-specific concerns might have been obscured or overlooked during analysis. A study by Richards (2005) highlighted that ICT-enabled lessons should not be divorced from the overarching goals of the curriculum. He also emphasized that the primary focus of ICT-enabled learning activities should be the 'curriculum focus' (such as the concepts and themes listed in the course syllabus), instead of the ICT tool itself. Therefore, school leaders and middle managers should find ways to help teachers feel more comfortable when using ICT to teach the core disciplinary understandings for the respective school subjects.

This chapter advances empirical research by focusing the discussion on lower secondary geography in Singapore to elaborate on how school leaders and middle managers support teachers in using ICT to enhance spatial thinking. In particular, this paper aims to:

1. establish greater understanding of how ICT is integrated into geography lessons that promote spatial thinking and inquiry and propose a taxonomy compartmentalizing the types of spatial thinking found in a typical ICT-enabled lower secondary geography lesson,
2. investigate the comfort level of teachers delivering these ICT-enabled lessons during geography classes,
3. elaborate on the support provided by school leaders and middle managers, and determine if such leadership is a strong variable that influences geography teachers' comfort level in using ICT, and
4. identify other factors that support teachers' comfort in using ICT to teach the lower secondary geography.

Clearly, such a research agenda was developed to provide baseline information on how ICT has been used to support learning. In-service geography teachers might find this information particularly useful because it helps them to assess, reflect and innovate some of their own teaching practices based in ICT. In addition, curriculum specialists—responsible for the planning and development of Geography Education in Singapore—might gain insight into how ICT is used in various schools to support geographical inquiry. This might be beneficial insofar as institutionalizing more informed ICT-initiatives to enrich the educative experiences of learners in the geography classroom. Agencies that cater to the professional development of geography teachers might also refer to these findings as a guide for designing ICT courses and seminars.

Further, this research study is expected to heighten our understanding of the technology culture of humanities departments in various secondary schools. The topic on how comfortable geography teachers are in using ICT tools for instruction is relatively neglected. Moreover, the specific ICT tools that they are comfortable with might have been overshadowed by earlier scholarship focusing on problematizing the technology integration in Singapore schools.

Finally, there is little discussion on the impact of leadership and middle management support on the 'community of practices' among geography teachers. While there has been significant understanding on the types of school leaders and middle managers, the roles they play and the areas of change they make to enhance technology integration in schools, little is known about how existing frameworks of leadership practices have influenced the 'set of tools and resources, approaches to teaching and learning, curriculum practices, cultural values, expectations and aims' among geography teachers (Hennessy et al. 2005, p. 160).

Factors Influencing Teachers' Use of ICT

A good number of studies have documented a multitude of factors influencing teachers' predisposition to use ICT in the classroom (e.g. Drent and Meelissen 2008; Cox et al. 2000; Haydn and Barton 2008; Buabeng-Andoh 2012; Mumtaz 2000). Surprisingly, Player-Koro (2012) showed that 'positive attitudes to ICT in pedagogical work with students' do not guarantee that teachers will use ICT in their lessons eventually (p. 105). In particular, Somekh's (2008) study of education systems in Europe has found that cultural, social and organizational contexts are key factors that determine the teachers' tendency to use ICT in the classroom. Somekh (2008) further theorized that teachers are not 'free agents', and their decision to deliver ICT-enabled lessons is symbiotically tied to 'regulatory frameworks and policies of national education systems and national cultures' as well as the 'student and local communities' (p. 450). This confirms studies by Afshari et al. (2009) and Vanderlinde et al. (2014) that teachers' use of ICT is dependent on highly complex interaction of factors.

Chigona and Chigona (2010) grouped these complex factors according to 'personal', 'social' and 'environmental' components. Tearle (2003a) suggests that two broad categories are essential in influencing teachers' use of ICT—the school environment and the teachers' professional development provisions. Clearly, a teacher's decision of ICT adoption is not solely based on an individual's preference. More importantly, these studies agree that 'school climate' is a powerful predictor of teachers' pedagogical adoption of ICT in lessons and educational leaders in the school are major stakeholders of determining the school climate. (See also Tezci 2011; Sang et al. 2009; Afshari et al. 2009). Divaharan (2007) employed a detailed questionnaire to study how the socio-cultural context of the school (such as the role of leadership and technical support) determines the type of ICT used in classrooms. Other studies also examined how school leaders are in the best position to articulate a shared ICT scheme that ensures the adequacy of support for teachers (Tondeur et al. 2008) and enables ICT to be used creatively for teaching and learning purposes (Scrimshaw 2004). However, there is still the issue of what using ICT in the classroom means. The notion of ICT integration will be examined in the next section.

Defining 'ICT Integration' and Teachers' 'Comfort Level' in Using ICT

Divaharan and Lim (2010) found that ICT integration solely refers to the 'manner in which the computer is used in the classroom during the course of lesson' in their study. On the other hand, Wang (2008) seeks to provide an expanded conception of 'ICT integration' by underlining the selection of the correct ICT tools 'for particular learning objectives or contexts' and the adaptation of existing resources to 'engage specific groups of learners' (p. 411). While the common parlance of 'ICT integration' is about how educational technologies are assimilated within the classroom learning environment, its definition has been deeply debated by academics. Wang's (2008) definition of ICT integration has been adopted in this study for two reasons. Although computers have been an essential part of the geography classroom, they are only one of the many ICT tools that are currently used to engage learners. With the proliferation of new smart devices designed with adroit telecommunicating capabilities to enhance geography learning, it would be more appropriate to examine ICT tools for the full range of functionalities, instead of merely the 'computer'. Second, Wang (2008) offers a more progressive definition of 'ICT integration' because it directs our perennial concerns of ICT use to learning rather than the technology itself. Indeed, ICT integration should be concerned with how its use will transform the learning process rather than evaluating how good a tool it is (Hayes 2007). Not all ICT tools are effective in mediating learning, and so Wang's (2008) definition is used because it encapsulates learning as the requisite for successful 'ICT integration'.

'Comfort level' is also a highly subjective term that requires more clarification in this paper as its meaning varies across industries and among professionals. From the perspective of inpatient care, a physician's 'comfort level' in performing a particular procedure is determined by his/her level of confidence to manage the complications that could arise (Huang et al. 2006). From an international marketing relationship perspective, a firm's 'comfort level' with a business partner involves 'trust, confidence and reciprocity of effort' (Witkowski and Thibodeau 1999). The notion of 'comfort level' is unique to each respective field. Nevertheless, both examples reflect 'comfort' as being familiar or at ease with the subject, as a key indicative emotion. This notion applies to the use of ICT for teaching and learning as the study by Papanastasiou and Angeli (2008) sought to outline five emotional/behavioural attributes of teachers demonstrating a high comfort level in using computers during lessons:

1. Teachers are not afraid of delivering computer-enabled lessons
2. Teachers do not feel anxious or tense delivering computer-enabled lessons

3. Teachers are prepared to troubleshoot should they encounter any technical issues
4. Teachers are receptive to the idea of computer as a tool for student engagement
5. Teachers are convinced that computer is a helpful tool to enhance teaching and learning.

Another study by Teo (2008) also proposed other attributes to gauge the perceived comfort level of teachers. For example, teachers would be able to 'teach [themselves] most of the things [they] need to know about computers', and teachers can 'make the computer do what [they] want it to' (p. 416). This understanding of 'comfort level' is pivotal for this study and it also allows the researcher to provide clarity to the subjects during data collection.

School Leadership in ICT Implementation

There is not yet a scholarly consensus as to how 'leadership' can bring about effective and successful ICT reform in the school (Leithwood and Duke 1999). School leaders certainly play a major technological role in the broad aspects of 'infrastructure, organization structure and policy, pedagogy and learning and school culture' (Tan 2010, p. 896). Yuen et al. (2003) posit that the 'vision and understanding of the role and impact of ICT in the curriculum', as well as the 'goals and objectives for ICT integration' of school leaders (p. 158), impact considerably on the success of technological reform. More precisely, they are responsible for articulating a set of comprehensive and achievable ICT goals for the school (see Yee 2000; Hadjithoma-Garstka 2011; Voogt et al. 2013 for reviews of the literature). This clarity in articulation is necessary for school leaders to develop a shared and common vision of ICT adoption with teachers who are ultimately implementing these technology plans in school (Buabeng-Andoh 2012).

In an article entitled 'A century's quest to understand school leadership', Leithwood and Duke (1999) outline six types of leadership practices of schools:

1. Instructional leadership, where the decisions of leaders will have a direct impact on teachers' behaviour as well as students' learning.
2. Transformational leadership, where leaders are dedicated to ensuring that their teachers are empowered to achieve greater goals.
3. Moral leadership, where the decisions of the leaders are made based on the leaders' own values or principles.
4. Participative leadership, where teachers gather together to contribute to the decision-making process, instead of solely the role of leaders in determining the final outcome.
5. Managerial leadership, where the decisions of leaders are made based on ensuring organizational efficiency.

6. Contingent leadership, where leaders' responses are adapted to the nature of the problem they face.

In the case of Singapore, transformational leadership was perceived to be helpful in dealing with the challenges involved in the school's ICT integration (Ng 2008), as well as increasing the level of sophistication of ICT integration (Ong and Tan 2011). Although Chen (2013) posits that transformational and instructional leaderships are the most common leadership functions found in a school's ICT implementation, Ng and Ho (2012) argue that the leadership practices of senior management are more transformational in nature, while those of the middle management are more instructional in nature. Ultimately, as Chen (2013) also argues, the heads of technology are seen to be performing these leadership roles more than the other school leaders and middle managers.

The previous section has attempted to unpack the complex and multifaceted concept of 'leadership' from works in the literature in multiple ways. While school leaders and middle managers are unambiguously central figures in determining the success, rate and scale of technological integration (Flanagan and Jacobsen 2003), there is insufficient clarity as to whether this is consistent across all levels of education and subjects. Research tends to be generic about the performances of leadership in ICT implementation, and what is surprisingly missing is a developed analysis of how it shapes the modes of respective subject teaching and learning. Therefore, this chapter examines these apparent interstices of scholarship, by presenting the perspectives of Singapore teachers' specializing in the teaching of lower secondary secondary.

Emphasis of Twenty-First Century Geography Education in Singapore

One of the major developments in the lower secondary geography syllabus in Singapore is that it embraces an issue-based approach, where students examine the 'significant environmental and human issues confronting Singapore and the world'. (MOE 2014) Through real-world examples and case studies, the intention is for students to be equipped with important spatial thinking skills. According to a report by the National Research Council (NRC) based in Washington, D.C., spatial thinking encompasses the following three cognitive competencies, 'concepts of space, tools of representation and reasoning processes' (NRC 2006). Jo and Bednarz (2009) provided a more developed analysis of these terms: 'Concepts of space' is characterized as the 'building blocks of spatial thinking', and examples include 'location, distribution, region, pattern, distance decay and spatial association'. 'Tools of representation' refer to any diagrammatic representations (such as maps and charts) that help to 'organize, understand and communicate information'. Finally, 'reasoning processes' refers to combined tasks of 'gathering information, processing information and generating new knowledge'. These findings show that the process of 'spatial thinking' is intended to model the work of a professional geographer among the lower secondary school students and help them appreciate the 'interconnectedness' of Singapore with other places in the world, i.e. 'what happens elsewhere can impact Singapore' (MOE 2013, p. 20).

Another new element embedded in the lower secondary geography syllabus is the call for 'geographical inquiry' as the recommended approach to the teaching and learning of geography. 'Geographical inquiry' is a pedagogic approach where students 'ask relevant questions, to pose and define problems, to plan what to do and how to research, to predict outcomes and anticipate consequences, and to test conclusions and improve ideas' (Roberts 2003, as cited in MOE 2014) The contention is that it 'encourages questioning, investigation and critical thinking about issues affecting the environment and people's lives, now and in the future' (MOE 2013, p. 29). Sorensen's (2009) literature review of Australia's National geography curriculum (as cited in Lupton 2012) also supported this view. The author pointed out that geographical inquiry 'incorporates the analytical, critical and speculative methodologies from the humanities as students of geography examine the impact of space, place and systems on the human condition' (p. 12). Yet, Taylor, Richards, and Morgan (2015) also warned that geographical inquiry is not tantamount to an extended geographical research task. The authors added that geographical inquiry can take place through any day-to-day lesson activities, and as such, '[g]eographical inquiry is as much an approach to pedagogy as it is a product'.

While the geography syllabus in Singapore has undergone numerous revisions since its inception during the 1970s (Chang 2012), the geographical concept of space underpinning the latest lower secondary geography syllabus (gazette by MOE in 2014) remained unchanged from its predecessor (gazette by MOE in 2006). This is not a surprise since the concept '[provides] valuable insights into the nature of geography' and '[anchors] the subject by giving it a greater coherence' (MOE 2014, p. 9). A number of academics have argued that geographical concepts like space behave as 'disciplinary grammars'—useful for bridging the 'link between the school subject and the academic discipline' (See Brooks 2013 for reviews of literature). Others like Vankan (2003) viewed such geographical concepts as 'instruments' necessary for students to make sense of the complex physical and human phenomena. Regardless of the differences in perceptions, geographical concepts positively influence students' understanding of geographical patterns and interactions (Lane 2008) and are imperative to a disciplined study of geography (Roberts 2014).

Using ICT to Support the Learning Goals of Geography

Hassell (2016) suggested five types of ICT tools commonly used by geography teachers—'presentation packages, data logging, data handling, simulations and modelling software and mapping and geographic information systems (GISs)'. The extensive array of ICT tools in geography is encouraging, and geography students will find these tools potentially valuable, especially in terms of 'testing hypotheses' and 'solving problems of an environmental and spatial nature' (Yeung 2010, p. 173)

Attitudes towards the use of ICT in supporting geographic education have been mixed. According to John and Sutherland (2004), the use of ICT in geography lessons has motivated students and clarified their understanding. Another scientific study by Demirci (2008) exploring the use of geographic information systems (GISs) in schools found that GIS enhances the visual elements of geography lessons and students demonstrated a higher level of competency in performing spatial analyses. However, Gregory (1981) (as cited in Morgan and Tidmarsh 2004) argued that ICT does not 'guarantee that the questions they were asking were any more meaningful or that the answers to them were any more incisive' (p. 191). Kent (2003) also contended that ICT-enabled geography lessons 'have been more ICT than geography-oriented'. In other words, the fact of the matter is that ICT on its own does not enrich geographical learning, but 'what it is used for that is most important' (Hassell 2016, p. 81).

Hence, as Morgan and Tidmarsh (2004) prescribe, ICT will need to be 'firmly based in an understanding of the aims and purposes of geography teaching'. In recent years, emerging research has drawn attention to the need to develop teachers' technological, pedagogical and content knowledge (TPACK). (See Angeli and Valanides 2008, 2009; Brupbacher and Wilson 2009) Coined by Mishra and Koehler (2006), TPACK is defined as the 'pedagogical techniques that use technologies in constructive ways to teach content' (p. 1029). One study by Doering, Koseoglu, Scharber, Henrickson, and Lanegran (2014) exploring the intersections of TPACK and geography Education in the USA found that the underpinning conceptual framework of TPACK proved useful in ensuring that geography teachers 'experiment with technological integration' that meets curricular objectives (p. 234). One way of equipping teachers with essential pedagogical knowledge of using various online resources and geospatial technologies to enhance geography learning in classrooms is through professional development programmes. While geography teachers responded that attending these courses built their confidence and motivation to apply the learned TPACK skill sets in their classes, these perceived benefits are also simultaneously overshadowed by broader institutional flaws such as 'limited technical support and infrastructure' (p. 234). (See also Bingimlas 2009; Jones 2004; Al-Senaidi et al. 2009; Salehi and Salehi 2012)

In the case of Singapore, research about geography teachers' use of ICT in geography has been relatively piecemeal. Apart from Chang (2002) and Sim, Lee, Chang, and Kho (2004) who studied the use of WebQuest to enhance geography learning among school students, most research has focused on the use of GIS in secondary schools. (See Seong 1996; Yap et al. 2008; Liu and Zhu 2008; Liu et al. 2012) Moreover, given that partner schools have only implemented GIS for the upper-secondary students, there is evidently a lack of understanding of the use of ICT in the teaching of lower secondary geography. Therefore, this research hopes to provide sorely needed insights at this level.

Methodology

The researchers interviewed ten in-service teachers specializing in the teaching of lower secondary geography. Each interview, which lasted approximately 45–60 min, was made up of the following two parts:

(a) Participation in a 15 min Likert item survey: Each subject rated their comfort level in using ICT and the involvement level of school leaders and middle managers in supporting a stronger technology culture.
(b) Participation in a 30–45 min conversation: Each subject shared more about how ICT has been used in the teaching of lower secondary geography to enhance geographical inquiry and whether he/she was comfortable in delivering these ICT-enabled lessons. The subject also gave details of the approach undertaken by school leaders and middle managers to nurture a strong technology culture and shared how this had influenced their comfort level in using ICT.

As earlier mentioned in the literature review, Divaharan (2007) offered a detailed study of how the socio-cultural environment of schools influenced teachers' type of ICT use in classrooms. Using 'Activity Theory' as a framework, part of her work included an investigation of how multiple levels of leadership changed the way teachers engaged students with the use of ICT. Since this study also examined the performances of leadership (but in shaping geography teachers' comfort level in using ICT), all survey items and conversation questions in this research were adapted from her questionnaire and modified to fit the context of this chapter.

The participants were recruited from the researchers' own networks of geography teachers and were deployed in five secondary schools during the time of the study. Of the ten interviewed, two were holding the position of Head of Department (Humanities) and another two were holding the position of Subject Head (geography) in their respective schools. As this research focuses on how school leadership and middle management influence geography teachers' comfort level in using ICT, including perspectives from a spectrum of geography teachers holding various leadership appointments in the school was useful in offering depth to the findings.

The interviews were conducted individually with the subjects. Before data collection commenced, the research objectives and the procedure for data collection were explained. The interview began with the definition of the boundaries of the research and a short clarification of the terms to be used. In particular, the definition of 'ICT integration' using Wang's (2008) study, and also the definition of 'comfort level' using Papanastasiou and Angeli's (2008) study were shared with the participants. Once the subjects understood the foci of the research, they proceeded to Stage 1 of the data collection, which encompassed four survey questions to rate (i) their personal comfort level in using ICT in the teaching of lower secondary geography, (ii) the involvement level of various stakeholders such as the HOD/ICT Head/Principal/Vice-principal in supporting a stronger technology culture,

(iii) modifications to comfort level attributed to the involvement of school leadership and middle management. In Stage 2 of the data collection, the conservation allowed the researchers to find out more about how geography teachers used ICT in lessons and how school leaders and middle managers influenced teachers' comfort level in pursuing ICT-incorporated instruction.

Guided by 11 questions, the conversation in Stage 2 also gave the participants an avenue to explain and clarify their earlier responses. This allowed the researchers to gain insights into the governance of ICT integration in schools. It was also an aim of this conversation to find out other factors that contributed to geography teachers' comfort level in using ICT.

Although the approach undertaken in this research was primarily qualitative, the Likert item survey was created for two reasons. First, through the survey, the researchers could have some understanding of participant's attitude and perception towards the use of ICT before the conversation with the participant began. This was helpful in setting the context of the subsequent conversation. Second, at the analysis stage of this study, it the researchers could corroborate the conversation with results from the survey.

The survey data was represented using charts, while the data from the conversations categorized and structured according to the four overarching themes of this paper, that are:

1. Types of spatial thinking found in a typical ICT-enabled lower secondary geography lesson.
2. Factors supporting geography teachers in using ICT to promote spatial thinking among lower secondary students. In particular, some of the reasons behind why geography teachers are comfortable/not comfortable delivering these ICT-enabled lessons were sifted for analysis.
3. Ways in which school leaders and middle managers provide support to enhance teachers' comfort level in using ICT and how these enhance teachers' comfort level in delivering only specific type(s) of spatial thinking.
4. Other factors that influence teachers' comfort level in using ICT to teach lower secondary geography.

These overarching themes also provide the analytical framework for further discussion.

Findings and Discussion

Types of ICT-Enabled Spatial Thinking Found in a Typical lower Secondary Geography Lesson

During the interviews, geography teachers were asked what ICT tools were used to impart geographical concepts of space in their lessons. Apart from geography

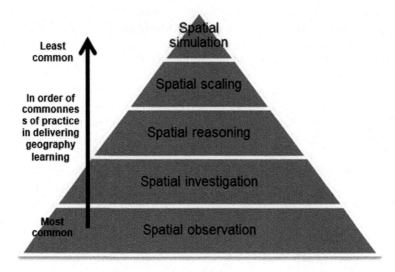

Fig. 12.1 Types of geography learning found in a typical ICT-enabled lower secondary geography lesson

lessons where teachers connected their electronic devices to an overhead projector for presentation-based lessons, students also gained access to ICT tools when they were brought to the computer lab, or when the teacher arranged 'mobile carts' (where the technical assistant of the school sets up laptops for every student in the class). One of the geography teachers also highlighted that all lower secondary students in her school owned a 'Chromebook'. Some students had access to smart devices such as iPads and GPS receivers. Together, we found that there were five types of spatial thinking found in a typical ICT-enabled lower secondary geography lesson ('Spatial observation', 'spatial investigation', 'spatial reasoning', 'spatial scaling' and 'spatial simulation'). This information is organized into a taxonomy based on how common each practice was among geography teachers (see Fig. 12.1).

Spatial Observation

'Spatial observation' is the most prevalent type of spatial thinking found in a lower secondary ICT-enabled geography lesson. Teachers used colours, sounds and moving images screened on the ICT device to elicit students' sense of sight and sound and impart various geographical concepts and ideas. 'Spatial observation' can be further sub-divided into two classes: teacher-led or student-led observations. For teacher-led spatial observations, the teacher typically planned and organized a mix of PowerPoint presentations and videos that were shown during lessons. Given that it was a highly structured and transmissive approach, students were akin to an 'audience'. In sum, most students were passive 'viewers' and 'listeners' in this context. One teacher shared how she used Google Earth to deliver a lesson on spatial observation:

> We taught housing with that. So we look at different countries, the different kind of settlement patterns, things like that, so we use Google Earth for that.
>
> – Teacher 4, School B

The same teacher also highlighted the importance of delivering an ICT-enabled lesson on spatial observation to enhance spatial understandings:

> No matter how much I tell them about place, space scale, they won't be able to know unless they see it for themselves. So sometimes, with the use of videos, sometimes even Google Earth, I will get them to see how big an area is, how big a country is, or how big a settlement is. So, a lot of moving images, basically, for them to be able to visualize.
>
> – Teacher 4, School B

For student-led spatial observations, the geography teacher brought students to the computer lab and assigned students with a prescribed list of online resources to view. Students had the autonomy to choose the resources they were interested in and decide how much time they would give themselves to complete the task during class:

> For example I want to teach effects of disasters like tsunami, so we will bring them to the Google Earth site where they will show the pre and post. You take back to the history of the area, so you can see before the tsunami, what had happened, what was the satellite picture like. Then after the tsunami, what the satellite pictures look like. Then the student can compare the extent of damage of the disaster. So in the sense they see the place, exact location of the event, then the scale of what they can see from the Google Earth, how extensive it is.
>
> – Teacher 5, School B

Spatial Investigation

'Spatial investigation' was another type of geography learning found in a lower secondary ICT-enabled geography lesson. As opposed to spatial observation where students undertook geographical inquiry solely by viewing and listening, 'spatial investigation' is more student-centred, and the task required students to acquire knowledge independently so as to fill in the missing information. However, instead of referring to the textbook, students were expected to accomplish the task by searching online sources for more information about the locational characteristics and spatial associations with the surrounding region, inquiring about what they had read and selecting what was necessary for them to answer the question. This typically took place in a computer laboratory where the teacher assigned students with questions about a topic that they had just learnt:

> We have to use Google Earth of where they are gonna walk, so they have to do research of the place, so they have to go to Thomson, and have a walk around the area to see the landuse, the heritage building, so there is a tour, and they will come back and do their land-planning.
>
> – Teacher 10, School E

As students were given the autonomy to choose what country they were interested in, they developed greater interest to find out more about spatial characteristics of the area, making this investigative approach useful for them to appreciate spatial patterns and distributions:

> Asking them to go and research a given city about population and housing, I think the students were interested to find more at least where it is located, a bit more of the city....
>
> – Teacher 7, School C

Spatial Reasoning

'Spatial reasoning' was the third type of geography learning found in a lower secondary ICT-enabled geography lesson. Similar to 'spatial investigation', students were assigned a series of tasks. However, spatial reasoning is more demanding, because students require higher-order cognitive skills in order to perform the task. In particular, they were expected to make sense of what they had read, and analyse and evaluate from different geographical contexts and case studies. Based on their findings, they were expected to present a logical and informed deduction of spatial issues using a mind map, poster, a report or a presentation slide:

> Sometimes we do create basic, what we call, IT worksheet to get students to think about. After that they watch the videos, what are their thoughts, because some of the syllabus requires them to think about their views, some things that they need to know, some things they can find out, some solutions they can implement.
>
> – Teacher 6, School B

Given the emergence of new online spaces such as Google Documents and Google PowerPoint which allow more than one user to create and edit documents simultaneously, many of the activities that encompassed spatial reasoning were also done in groups to enhance collaborative learning among students:

> Okay, so last year, we did rainforest in sec 1. There was not a lot of - so I use ICT for the analysis part - which is when they are collected all the data using pen and paper, they added all this into the Google Sheets, and whatever they have. I will use that as a basis to do a simple analysis.
>
> – Teacher 9, School D

Spatial Scaling

'Spatial scaling' was the fourth type of spatial thinking found in a lower secondary ICT-enabled geography lesson. The idea of 'spatial scaling' was developed through the appropriation of Christopher D. Lloyd's notion of 'spatial scale' (Lloyd 2014). In his book, he coined this term to describe 'the size or extent of a process, phenomenon or investigation'. O'Sullivan (2015) subsequently offered a more developed conceptualization of this term, by denoting that it is 'a strictly spatial data-oriented concept, intimately bound up with any attempt to characterize the geographical structure of social or physical phenomena'. For this reason, this chapter also applies the term 'spatial scaling' to encompass the spatial analytic methods used to '[understand] the implications of spatial scale for geographical data' (O'Sullivan 2015). Examples of this type of spatial thinking were scarcely found in the teachers' discussions. Most teachers discussed the representation of data through the use of graphical tools, but only a few discussed the spatial scaling of data through the use of mapping tools, for instance.

Spatial Simulation

'Spatial simulation' was the fifth, and the least common type of spatial thinking found in a lower secondary ICT-enabled geography lesson. According to O'Sullivan and Perry (2013), this term is defined as the process in which 'specific systems and phenomena' is represented, and it is one type of geography learning used by numerous academics to predict future patterns of change in an area over time (See studies by Turner 1988; Gustafson et al. 2000; and Fang et al. 2005). The focus on such geography learning is for students to experiment with ICT tools to simulate real-world processes and interactions. One of the teachers attempted to use online websites for students to recreate the process of water cycle through animation:

> They do a research on the water cycle, so we said; use videos or something else, and quite a number did StopMotion videos on their own. We give them a link on how to do StopMotion, so they will watch it and then they came up with their own ones.
>
> – Teacher 10, School E

Apart from using animations, there were other ICT tools for students to recreate a particular spatial region. The same geography teacher also incorporated the use of online applications for students to outline the spatial functions in a neighbourhood:

> Like what I have just told you, Google Earth, there's one Google 'My Maps'. So I got my lower sec to create their own Google map of a neighbourhood, let's say Tiong Bahru. That's what was done recently
>
> – Teacher 5, School B

Teachers' Comfort Level in Delivering ICT-Enabled Lessons During Geography Classes

While the previous section outlined the types of geography learning found in a typical ICT-enabled lower secondary geography lesson, the diagram below provides a background understanding of teachers' existing comfort level in delivering these lessons.

Based on the survey results shown in Fig. 12.2, all the geography teachers had a neutral or positive attitude towards the use of ICT in lessons and were comfortable with using ICT to deliver geography lessons. In particular, more than 50% of respondents felt comfortable delivering 'basic ICT-incorporated instruction during geography lessons', with three respondents who were comfortable selecting a variety of ICT tools and resources to conduct a geography class, and one respondent who was comfortable innovating its ICT resources to conduct 'an effective geography lesson'. Geography teachers were generally comfortable using ICT because of its efficacy in achieving curriculum goals:

> I suppose in the past when we didn't rely much on ICT in our lesson, now maybe because of the shift in curriculum, we tend to use more in ICT.
>
> – Teacher 4, School B

Furthermore, geography teachers felt encouraged to use ICT because students were more engaged during their lessons:

> At least when I bring them to the computer lab to work on 'My Maps', I won't say all, I won't say 100%, at least about 70% are more focused
>
> – Teacher 5, School B

One of the geography teachers shared that she was comfortable using ICT to deliver a lesson on spatial investigations, because students enjoyed what they had

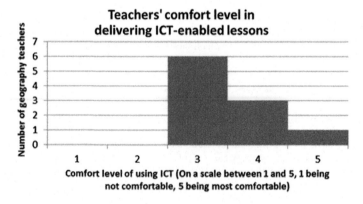

Fig. 12.2 Survey results of teachers' comfort level in delivering ICT-enabled lessons

learnt and she was satisfied that students' final submission of their class activity in the computer laboratory went beyond her expectations.

> Actually when they did up, it was not bad, the brief write-up that they did and they took pride of their work, I think that is something.
>
> – Teacher 7, School C

However, the above findings were caveated by the finding that most teachers were only comfortable delivering ICT-enabled lessons that involved 'spatial observation'. In particular, most geography teachers were inclined to conduct a teacher-led spatial observation lesson. As one of the geography teachers remarked:

> Only when we want to give them paper-based assignments, that we don't use the ICT tools. If not, it might be the usual teacher-fronted lesson, and is usually we have ICT tools to back us up.
>
> – Teacher 4, School B

Teachers also indicated that were not comfortable to use ICT tools beyond PowerPoint slides and videos regularly.

> I am definitely not an expert in ICT; I am definitely not very comfortable also. But I don't mind if let's say there is some kind of guidance to try along. But it won't be an everyday affair. You know, in every of my lesson to use ICT - that I don't think I will be very comfortable also.
>
> – Teacher 2, School A

Roles of School Leaders and Middle Managers in Developing a School-Wide ICT Implementation

Teachers also suggested that school principals and vice-principals played an important role in shaping the technology culture of the school and in supporting teachers in using ICT in their lessons This is similar to other studies by Tan (2010a) and Yuen et al. (2003), who found that they are responsible for cultivating a cohesive ICT environment where teachers are able to seek necessary technical support when required. As emphasized by one of the geography teachers about the role of school leaders in determining the overall direction of ICT implementation:

> I guess they give the encouragement and support. Because it comes from them, so if their direction is 'please use ICT as much as possible and try to interact with kids using that', I think it comes down to us, it cascades down to us also, so their support is very important. In case if we, let's say example, a new ICT to be implemented, we always know we have their support, and they will be able to help us in that.
>
> – Teacher 4, School B

Tearle (2003a) drew attention to the need for school leadership to encourage teachers to attend ICT training. The teachers in this study reported that their school leaders encouraged them to attend professional development courses and learn new pedagogical approaches in ICT.

> The support to give time off for these geography teachers to attend the workshops so that they can enrich their learning, if there are seminars that require funding, so that's where the principal or vice-principal give approval for funding, so that the geography teachers can then attend seminars, so that they will come back to share with the rest of us.

Despite the importance of school leaders in general in setting the culture of the school, and in making space and time for professional development, some of the participants also pointed to the more crucial role of middle managers in supporting ICT implementation. This was further supported by quantitative data indicating that the middle managers (subject head and head of department) were perceived as more directly involved in the school's ICT implementation than the principal and vice-principal (Fig. 12.3).

Notably, two teachers rated the involvement of principal and vice-principal as '2' (Does not make ICT-incorporated instruction a necessary effort for all geography lessons). This result is expected; since principals are not expert in the teaching of all subjects, and the school practised a distributed leadership in the ICT implementation. This also corresponds with one of the interviews with a geography teacher. When asked about how the principal and vice-principal were involved in the ICT implementation, she replied:

> This has been passed for quite some time already, so they are not as hands-on now. It is all passed down to the Heads or HODs, if they tell us that, it has been told or agreed by the Heads, so they do during the meeting they do tell us to try to use, or asks for updates of how ICT can be used to enhance teaching.
>
> – Teacher 10, School E

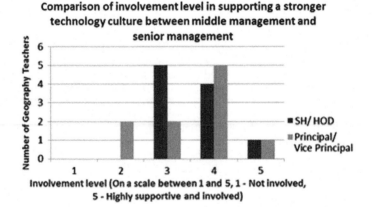

Fig. 12.3 Survey results depicting what geography teachers feel about the involvement level of school leaders/senior management and middle management

Department's ICT Goal in the Teaching of Lower Secondary Geography

Given that teachers perceive middle managers as more in shaping teachers' comfort level in using ICT, the aim of this section is to analyse the department's ICT goal in influencing the teaching of lower secondary geography. The ICT goals of geography teachers must be aligned with departmental goals in order for ICT implementation within the school to be effective. (Yuen et al. 2003). One of the geography teachers (who is assuming the role of the Head of Department for humanities) articulated a clear goal to all teachers specializing in the teaching of lower secondary geography:

> For the lower sec, their ICT plan would be to engage students; it would still be more of usage of YouTube videos, so that students are able to have a visual of the topic, besides the textbook. Then, researching information, because researching information would be important for GI. Especially for Sec 1 s, they might not know how to do research with information. So far, these are the two basic plans we have.
>
> – Teacher 6, School B

When geography teachers in the same school were interviewed, they also spelled out similar ICT goals as their HOD by underlining the need for lower secondary students to be competent in researching information. This showed that the teachers identified with the ICT goals of the leadership:

> Basically, I think ours is to equip students with the ICT knowledge as well. So it also helps them to be able to use ICT in, when they do their research or when they do their homework. So even though we give them worksheet and all, we hope that they will be able to use ICT tools to help them find answers, rather than relying on the textbook.
>
> – Teacher 4, School B

In this school, where the ICT goals of geography teachers were aligned with that of their middle manager, teachers were comfortable using a variety ICT tools during their lessons.

> Things like using Google Earth to chart up a direction during their tourism fieldwork, or even how to calculate using weather instrument so on and so forth, so we try to incorporate that right now.
>
> – Teacher 4, School B

Not all departments had clear goals when it came to the use of ICT.

> The goals are not concrete; I think it is quite laissez-faire.
>
> – Teacher 9, School D

> Not that I know of. Ok, the only thing we need to do is basically in case there is a, like say, school closure, we must always have something prepared, (e-learning), so it could be.
>
> – Teacher 2, School A

In these schools, where the geography department did not communicate any specific ICT goals, geography teachers did not always feel comfortable incorporating ICT-based learning in their lessons, particularly if they were time-consuming.

> I had a class last time; I taught them map-reading. So I made use of online games, to help them to identify. So it is like a game, you have to escape from the wolf or something for sec 1… but you use up time; students are just playing and you waste their energy level.
>
> – Teacher 1, School A

Types of Leadership: ICT Implementation of Lower Secondary Geography

A study by Leithwood and Duke (1999) found that ICT implementation in a humanities department can be typified by one of the three types of leadership practices: managerial leadership, instructional leadership and transformational leadership. Such performances of leadership influence the type of ICT-enabled geography learning geography teachers are comfortable in.

Managerial leadership is common in schools where decisions of leaders are made based on ensuring organizational efficiency. This is exemplified in how school leaders impose a minimum percentage of lessons to be delivered in ICT and a minimum number of ICT courses teachers must attend in a single school year to enable an efficient process of school-wide ICT implementation:

> It is compulsory actually, it is compulsory in the sense that you will have to sign up, as in you definitely have to sign up for all, just that different timing. That's all. But we attend.
>
> — Teacher 2, School A

> ICT goal, I would say my target is 50% of all lesson, to have all ICT-based.
>
> – Teacher 8, School C

Yet, geography teachers who are operating under this leadership model also added that most of their ICT-enabled lessons only encompass teacher-centred 'spatial observation'. This is expected because geography teachers might just want to fulfil the minimum expectations required of them. Teachers—who do not see how other types of ICT-enabled geography learning can enhance the geographical inquiry of students—would simply achieve the minimum key performance index (KPI) by replicating the same teaching process that was done previously, i.e. presentation slides and videos.

Instructional leadership (where the decisions of leaders will have a direct impact on teachers' behaviour as well as students' learning) is another common leadership practice in the humanities department. This is demonstrated in how the head of department expects all geography teachers in the department to collaborate and achieve an ICT-based project together:

> The humanities department, every teacher must video, go into a room and video, so let's say for geography, the steps involved in arriving at an answer, so that you can actually play in class when the teacher is absent, can play and revise, so each teacher, we use ICT, we do videos last year. I do one video; the teacher [does] another video, every teacher minimum do 1 video, so we have a collection of resources of video.
>
> – Teacher 3, School A

However, these videos are mainly used for contingency plans. Even though it can be incorporated into their lessons, the final ICT product that was done collaboratively as a department does little to incentivize geography teachers to give up their standard teaching practices which they have become comfortable with.

On the other hand, in transformational leadership, where leaders are dedicated to ensuring that their teachers are empowered to achieve greater goals, ICT use proved more beneficial for geography teachers. This is reflected in how one of the schools assembled a group of passionate geography teachers' specialized in developing new geography lesson packages that use ICT to encompass different types of geography learning and sharing with the rest of the department:

> This group of people is there to help us, basically champion the whole ICT move, and help us to infuse as much ICT in our lesson as possible…
>
> – Teacher 4, School B

This is appreciated by other geography teachers, who believe that sharing a ready pool of resources that have been produced (and even tested) by other teachers helps them feel more confident and comfortable using it with their own class:

> But you don't know our subject areas right, are the new technologies, somebody needs to update us to support learning. Because we are not expert in that. What we want is immediately of use that can make our lesson interesting, but ultimately is like, you must give us something to experiment with them and actually implement.
>
> – Teacher 3, School A

Other Factors Influencing Teachers' Comfort Level in Using ICT to Teach Lower Secondary Geography

One of the other factors influencing geography teachers' comfort level in using ICT includes the types of ICT tools teachers use to deliver the geography learning. In particular, some ICT tools may pose privacy concerns for the school. As mentioned during the interview:

> They introduce you to all freemiums like free stuff like Edmodo, but we have been advised by HOD ICT not to use it because it requires personal data and privacy issues, they said it is like too… too problematic, so they discourage people from using this kind of freemiums.
>
> – Teacher 9, School D

The same geography teacher also highlighted the nature of working environment —where geography teachers are situated—as one significant factor influencing his comfort level. In particular, it was important for the department to have a culture of sharing among the teachers:

> When they tried it like they share it with their colleagues this thing right, and then some of them might be curious to try out. But if nobody tells anyone or nobody shares, then I think nobody would think that such a thing could be used.
>
> – Teacher 9, School D

The critical importance of the TPACK was also accentuated when one geography teacher suggested having more ICT tools and resources that are relevant to the teaching of the geography syllabus:

> Content must be suitable. Because sometimes when you talk about distribution then it is very hard to talk about… Like the distribution of water supply in the world - then actually this is very hard to do ICT. Because all these are very… I mean it depends on the content. Certain content will allow you to use OCT.
>
> – Teacher 1, School A

In addition, a geography teacher argued that the comfort level cannot be easily modified. The school management can only do so much for geography teachers. However, when it comes to the final lesson delivery, it depended on teachers' attitude towards the use of ICT. In his words, he asserted that:

> It really depends on the teacher. If the teacher doesn't feel or doesn't buy the idea, and is not confident of delivering content in such a manner, then the students will walk out with lesser confidence in whatever they have learnt and the teacher. So comfort level is not easily changed.
>
> – Teacher 8, School C

Conclusion

Ultimately, leadership plays an important role in influencing the type of ICT-enabled geography learning geography teachers are comfortable to deliver. In particular, middle managers play a more instrumental role than senior managers in enhancing their teachers' comfort level in using ICT. Moreover, this study found that there are five types of geography learning found in a typical ICT-enabled lower secondary geography lesson: 'Spatial observation', 'spatial investigation', 'spatial reasoning', 'spatial scaling' and 'spatial simulation'. The findings show that transformational leadership in the humanities department is most beneficial to teachers specializing in the teaching of lower secondary geography because it helped them to be more comfortable using ICT tools to conduct a variety of geography learning. On the other hand, a managerial or an instructional leadership in the humanities department produced only teachers who are comfortable delivering ICT-enabled lessons that involve 'spatial observation'.

Three broad recommendations are proposed based on this study. First, middle managers must recognize the critical importance of their role in shaping a cohesive and sharing ICT culture among geography teacher. Middle managers like the Head of Department for humanities or the Subject Head for geography must communicate a clear ICT goal to the teachers. This study also proposes that middle managers adopt a 'transformational leadership' in managing the ICT implementation of the school, instead of a 'managerial' or 'instructional' leadership.

Second, senior managers of the school (such as the principal and vice-principal) might not be subject matter experts in geography. However, they can provide the necessary financial support in ensuring that the adequacy, efficiency and reliability of the school's ICT infrastructure are continuously enhanced.

Third, professional development courses might better cater to the needs of geography teachers if they equip teachers with ICT tools that are relevant to the geography syllabus. One suggestion is to align the course outline with the TPACK framework (as proposed by Mishra and Koehler (2006)) so that teachers will be able to apply the same skills learnt during the course in the classroom. In considering the interaction between the teacher, the geography subject and the student experiences, the pedagogical content knowledge allows the teacher to interpret the curriculum documents and design learning activities for the student, but is at the same time guided by the subject disciplinary knowledge of geography that will drive the pedagogy and choice of technology use as it requires the teacher to carefully consider what learning activity to choose, what the key concepts are and how they can help students think geographically to take them beyond what they already know.

To understand how school leadership and middle management impact the geography teachers' comfort level in using ICT, this study included geography teachers who were also assuming the role of subject head or head of department. The intention is to provide a balanced perspective. However, one possible limitation of this study is that senior leadership (such as the principal and vice-principal) were not interviewed. While the heads of department involved in this study do provide a leadership perspective of this issue, the principal or vice-principal of the school might also have alternative views to share. Another limitation of this study is that geography learning is only confined to 'space'. Yet, the lower secondary geography syllabus in Singapore is also underpinned by other concepts, such as place, scale and physical and human processes. The authors hope that future research could also look into how ICT has been used by geography teachers to communicate and impart these concepts. More developed analysis of other factors that influence geography teachers' comfort level in using ICT can also be considered.

The learning revolution (Resnick 2002) that has been described as education mapped against new terrains of pedagogic practices has resulted in various stakeholders of the education system responding differently to the use of ICT in the classroom. While school leadership and middle management have clear influences on the adoption of ICT in instruction, the prevalence and type of ICT use in a geography classroom is further compounded by other factors. This has led to a two-part treatment of the discussion. While the divide is artificial and the two issues often exist simultaneously for a classroom teacher wanting to use ICT in the geography lesson, the analysis of the empirical research has allowed us to highlight the equivalence in importance of both issues. The authors remain hopeful by the sentiments shared by Teacher 8 of School C, that ultimately it is the confidence of the teacher, regardless of the support structures that will ensure the meaningful and successful delivery of an geography ICT lesson. And the authors argue that the confidence of teachers can be built on strong disciplinary knowledge of geography and continual professional development.

References

Afshari, M., Bakar, K. A., Luan, W. S., Samah, B. A., & Fooi, F. S. (2009). Factors affecting teachers' use of information and communication technology. *Online Submission, 2*(1), 77–104.

Al-Senaidi, S., Lin, L., & Poirot, J. (2009). Barriers to adopting technology for teaching and learning in Oman. *Computers & Education, 53*(3), 575–590.

Angeli, C., & Valanides, N. (2008, March). TPACK in pre-service teacher education: Preparing primary education students to teach with technology. In *AERA Annual Conference*, New York.

Angeli, C., & Valanides, N. (2009). Epistemological and methodological issues for the conceptualization, development, and assessment of ICT–TPACK: Advances in technological pedagogical content knowledge (TPACK). *Computers & Education, 52*(1), 154–168.

Bingimlas, K. A. (2009). Barriers to the successful integration of ICT in teaching and learning environments: A review of the literature. *Eurasia Journal of Mathematics, Science & Technology Education, 5*(3).

Brooks, C. (2013). How do we understand conceptual development in school geography. *Debates in Geography Education*, 75–88.

Brupbacher, L., & Wilson, D. (2009). Developing technological pedagogical content knowledge in teacher preparation programs. Paper presented at the *Proceedings of Society for Information Technology & Teacher Education International Conference*.

Buabeng-Andoh, C. (2012). Factors influencing teachers' adoption and integration of information and communication technology into teaching: A review of the literature. *International Journal of Education and Development using Information and Communication Technology, 8*(1), 136.

Chang, C. H. (2002). Designing web-based constructivist learning activities for geography and social studies.

Chang, C. H. (2012). A critical discourse of Singapore's school geography for the twenty-first century. *Literacy Information and Computer Education Journal, 3*(3), 628–638.

Chen, W. (2013). School leadership in ICT implementation: Perspectives from Singapore. *The Asia-Pacific Education Researcher, 22*(3), 301–311.

Chen, W., Tan, A., & Lim, C. (2012). Extrinsic and intrinsic barriers in the use of ICT in teaching: A comparative case study in Singapore. Paper presented at the *Australasian Society for Computers in Learning in Tertiary Education (ASCILITE) Conference*, New Zealand.

Chigona, A., & Chigona, W. (2010). An investigation of factors affecting the use of ICT for teaching in the Western Cape Schools. Paper presented at the *ECIS, South Africa*.

Cox, M. J., Cox, K., & Preston, C. (2000). What factors support or prevent teachers from using ICT in their classrooms?

Demirci, A. (2008). Evaluating the implementation and effectiveness of GIS-based application in secondary school geography lessons. *American Journal of Applied Sciences, 5*(3), 169–178.

Divaharan, S. (2007). Secondary school socio-cultural context influencing teachers' type of ICT use: a case study approach. (Doctor of Philosophy), Nanyang Technological University, Singapore.

Divaharan, S., & Lim, C. P. (2010). Secondary school socio-cultural context influencing ICT integration: A case study approach. *Australasian Journal of Educational Technology, 26*(6), 741–763.

Doering, A., Koseoglu, S., Scharber, C., Henrickson, J., & Lanegran, D. (2014). Technology integration in K–12 geography education using TPACK as a conceptual model. *Journal of Geography, 113*(6), 223–237.

Drent, M., & Meelissen, M. (2008). Which factors obstruct or stimulate teacher educators to use ICT innovatively? *Computers & Education, 51*(1), 187–199.

Fang, S., Gertner, G. Z., Sun, Z., & Anderson, A. A. (2005). The impact of interactions in spatial simulation of the dynamics of urban sprawl. *Landscape and urban planning, 73*(4), 294–306.

Flanagan, L., & Jacobsen, M. (2003). Technology leadership for the twenty-first century principal. *Journal of Educational Administration, 41*(2), 124–142.

Gregory, D. (1981). *Towards a human geography*. UK: Longman Harlow.

Gustafson, E. J., Shifley, S. R., Mladenoff, D. J., Nimerfro, K. K., & He, H. S. (2000). Spatial simulation of forest succession and timber harvesting using LANDIS. *Canadian Journal of Forest Research, 30*(1), 32–43.

Hadjithoma-Garstka, C. (2011). The role of the principal's leadership style in the implementation of ICT policy. *British Journal of Educational Technology, 42*(2), 311–326.

Hassell, D. (2016). *Issues in geography teaching*. USA: Routledge.

Haydn, T., & Barton, R. (2008). 'First do no harm': Factors influencing teachers' ability and willingness to use ICT in their subject teaching. *Computers & Education, 51*(1), 439–447.

Hayes, D. N. (2007). ICT and learning: Lessons from Australian classrooms. *Computers & Education, 49*(2), 385–395.

Hennessy, S., Ruthven, K., & Brindley, S. (2005). Teacher perspectives on integrating ICT into subject teaching: commitment, constraints, caution, and change. *Journal of Curriculum Studies, 37*(2), 155–192.

Huang, G. C., Smith, C. C., Gordon, C. E., Feller-Kopman, D. J., Davis, R. B., Phillips, R. S., et al. (2006). Beyond the comfort zone: Residents assess their comfort performing inpatient medical procedures. *The American Journal of Medicine, 119*(1), 71. e17–71. e24.

Hung, D. W. L., Tan, S. C., Chong, D. P. Y., Wong, P. S. K., Cheah, H. M., Tan, H. C., et al. (2003). Projecting ICT developments in teaching and learning for the near future: restructuring the landscape of teaching and learning interactions. *Teaching and Learning, 24*(1), 97–104.

Jo, I., & Bednarz, S. W. (2009). Evaluating geography textbook questions from a spatial perspective: Using concepts of space, tools of representation, and cognitive processes to evaluate spatiality. *Journal of Geography, 108*(1), 4–13.

John, P. D., & Sutherland, R. (2004). Teaching and learning with ICT: New technology, new pedagogy? *Education, Communication & Information, 4*(1), 101–107.

Jones, A. (2004). A review of the research literature on barriers to the uptake of ICT by teachers.

Kent, W. A. (2003). *Geography and information and communications technologies: Some futures thinking International handbook on geographical education* (pp. 337–344). Springer.

Lane, R. (2008). Students' alternative conceptions in geography. *Geographical Education, 21*(1), 43.

Leithwood, K., & Duke, D. (1999). A century's quest to understand school leadership. *Handbook of research on educational administration* (Vol. 2, issue no. 1, pp. 45–72).

Liang, J.-C., Chai, C. S., Koh, J. H. L., Yang, C.-J., & Tsai, C.-C. (2013). Surveying in-service preschool teachers' technological pedagogical content knowledge. *Australasian Society for Computers in Learning in Tertiary Education, 29*(4), 581–594.

Liu, S., & Zhu, X. (2008). Designing a structured and interactive learning environment based on GIS for secondary geography education. *Journal of Geography, 107*(1), 12–19.

Liu, Y., Tan, G. C. I., & Xiang, X. (2012). Singapore: The information technology masterplan and the expansion of GIS for geography education International perspectives on teaching and learning with GIS in secondary schools (pp. 215–224): Springer.

Lloyd, C. D. (2014). *Exploring spatial scale in geography*. UK: Wiley.

Looi, C.-K., Hung, D. W. L., Bopry, J., & Koh, T. S. (2004). Singapore's learning sciences lab: Seeking transformations in ICT-enabled pedagogy. *Educational Technology Research and Development, 52*(4), 91–115.

Lupton, M. (2012). Inquiry skills in the Australian curriculum. *Access, 26*(2), 12.

Mishra, P., & Koehler, M. J. (2006). Technological pedagogical content knowledge: A framework for teacher knowledge. *Teachers College Record, 108*(6), 1017.

MOE. (2013). A guide to teaching lower secondary geography. Singapore.

MOE. (2014). Lower secondary geography teaching syllabuses. Singapore.

MOE. (2016). Educational Technology. Retrieved from https://www.moe.gov.sg/about/org-structure/etd.

Morgan, J., & Tidmarsh, C. (2004). Reconceptualising ICT in geography teaching. *Education, Communication & Information, 4*(1), 177–192.

Mumtaz, S. (2000). Factors affecting teachers' use of information and communications technology: A review of the literature. *Journal of information technology for teacher education, 9*(3), 319–342.

Ng, D., & Ho, J. (2012). Distributed leadership for ICT reform in Singapore. *Peabody Journal of Education, 87*(2), 235–252.

Ng, W. L. (2008). Transformational leadership and the integration of information and communications technology into teaching.

NRC. (2006). *Learning To Think Spatially*. Washington, D.C.: National Academy Press.

O'Sullivan, D., & Perry, G. L. (2013). *Spatial simulation: Exploring pattern and process*. New Zealand: Wiley.

O'Sullivan, D. (2015). Exploring spatial scale in geography. *International Journal of Geographical Information Science, 29*(10), 1932–1933.

Ong, A. K.-K., & Tan, S. C. (2011). Distributed leadership for integration of information and communication technology (ICT) in schools.

Papanastasiou, E. C., & Angeli, C. (2008). Evaluating the use of ICT in education: Psychometric properties of the survey of factors affecting teachers teaching with technology (SFA-T3). *Educational Technology & Society, 11*(1), 69–86.

Player-Koro, C. (2012). Factors influencing teachers' use of ICT in education. *Education Inquiry, 3*(1), 93–108.

Resnick, M. (2002). Rethinking learning in the digital age: The Global Information Technology Report: Readiness for the Networked World. Oxford University Press.

Richards, C. (2005). The design of effective ICT-supported learning activities: Exemplary models, changing requirements, and new possibilities. *Language Learning & Technology, 9*(1), 60–79.

Roberts, M. (2003). *Learning through enquiry: Making sense of geography in the key stage 3 classroom*. UK: Geographical Association.

Roberts, M. (2014). Powerful knowledge and geographical education. *Curriculum Journal, 25*(2), 187–209.

Salehi, H., & Salehi, Z. (2012). Integration of ICT in language teaching: Challenges and barriers. Paper presented at the *International Conference on e-Education, e-Business, e-Management and e-Learning, Singapore*.

Sang, G., Valcke, M., Van Braak, J., & Tondeur, J. (2009). Factors support or prevent teachers from integrating ICT into classroom teaching: A Chinese perspective. Paper presented at the *Proceedings of the 17th International Conference on Computers in Education [CDROM]* (pp. 808–815), Hong Kong: Asia-Pacific Society for Computers in Education. Retrieved from http://www.icce2009.ied.edu.hk/pdf/c6/proceedings808-815.pdf.

Scrimshaw, P. (2004). Enabling teachers to make successful use of ICT.

Seong, K. T. (1996). Interactive multimedia and GIS applications for teaching school geography.

Sim, H. H., Lee, C. K.-E., Chang, C. H., & Kho, E. M. (2004). Exploring the use of WebQuests in the learning of social studies content. *Teaching and Learning, 25*(2), 223–232.

Somekh, B. (2008). *Factors affecting teachers' pedagogical adoption of ICT International handbook of information technology in primary and secondary education* (pp. 449–460). USA: Springer.

Sorensen, L. (2009). Literature Review—For the National Geography Curriculum. *Geographical Education, 22*(1), 12–17.

Tan, S. C. (2010a). School technology leadership: Lessons from empirical research.

Tan, W. L. (2005). *Integrating ICT in the learning of recorders in the music classroom.* (Master of Education). Singapore: Nanyang Technological University.

Tan, Y. L. (2010). *Leading change in ICT: A case study of leadership styles.* (Master of Arts (Instructional design and technology)). Singapore: Nanyang Technological University.

Tay, L. C. (2014). *Content knowledge, pedagogy and technology in Singapore classrooms: an approach to integration and development.* (Doctor in Education). Singapore: Nanyang Technological University.

Tay, L. Y. (2002). *Using information and communication technology tools to engage students in higher-order thinking—A case study.* (Master of Arts (Instructional design and technology)). Singapore: Nanyang Technological University.

Taylor, M., Richards, L., & Morgan, J. (2015). *Geography in focus: Teaching and learning in issues-based classrooms.* New Zealand: NZCER Press.

Tearle, P. (2003a). Enabling teachers to use information and communications technology for teaching and learning through professional development: Influential factors. *Teacher Development, 7*(3), 457–472.

Tearle, P. (2003b). ICT implementation: What makes the difference? *British Journal of Educational Technology, 34*(5), 567–583.

Teo, T. (2008). Pre-service teachers' attitudes towards computer use: A Singapore survey. *Australasian Journal of Educational Technology, 24*(4), 413–424.

Tezci, E. (2011). Factors that influence pre-service teachers' ICT usage in education. *European Journal of Teacher Education, 34*(4), 483–499.

Toh, E. L. P. (2013). *Capabilities and preferences of incorporating ICT: A study of preschool teachers in Singapore.* (Master of Education). Singapore: Nanyang Technological University.

Tondeur, J., Valcke, M., & Van Braak, J. (2008). A multidimensional approach to determinants of computer use in primary education: Teacher and school characteristics. *Journal of Computer Assisted learning, 24*(6), 494–506.

Turner, M. G. (1988). A spatial simulation model of land use changes in a piedmont county in Georgia. *Applied Mathematics and Computation, 27*(1), 39–51.

Vanderlinde, R., Aesaert, K., & Van Braak, J. (2014). Institutionalised ICT use in primary education: A multilevel analysis. *Computers & Education, 72,* 1–10.

Vankan, L. (2003). Towards a new way of learning and teaching in geographical education. *International Research in Geographical and Environmental Education, 12*(1), 59–63.

Voogt, J., Knezek, G., Cox, M., Knezek, D., & ten Brummelhuis, A. (2013). Under which conditions does ICT have a positive effect on teaching and learning? A Call to Action. *Journal of Computer Assisted learning, 29*(1), 4–14.

Wang, Q. (2008). A generic model for guiding the integration of ICT into teaching and learning. *Innovations in Education and Teaching International, 45*(4), 411–419.

Witkowski, T. H., & Thibodeau, E. J. (1999). Personal bonding processes in international marketing relationships. *Journal of Business Research, 46*(3), 315–325.

Yap, L. Y., Ivy Tan, G. C., Zhu, X., & Wettasinghe, M. C. (2008). An assessment of the use of geographical information systems (GIS) in teaching geography in Singapore schools. *Journal of Geography, 107*(2), 52–60.

Yee, D. L. (2000). Images of school principals' information and communications technology leadership. *Journal of Information Technology for Teacher Education, 9*(3), 287–302.

Yeung, S. P.-M. (2010). IT and geography teaching in Hong Kong secondary schools: A critical review of possibilities, trends and implications. *International Research in Geographical and Environmental Education, 19*(3), 173–189.

Yuen, A. H., Law, N., & Wong, K. (2003). ICT implementation and school leadership: Case studies of ICT integration in teaching and learning. *Journal of Educational Administration, 41* (2), 158–170.

Noah Zhang is an aspiring geographer who was selected for the prestigious NTU-NIE Teaching Scholars Programme (TSP). His perspective is informed by experiences in secondary school contexts both in Singapore and his overseas school attachment at Aarhus, Denmark. He is interested to advance his work on the role technology and school leadership in geography education.

Tricia Seow is an experienced geography educator and involved in the Fieldwork Exercise Task Force of the International Geography Olympiad and in the MOE Humanities Talent Development Programme. She also serves as the Hon-Gen Secretary of the Southeast Asia Geography Association. She enjoys fieldwork, and her research interest includes teachers' knowledge and practice, field inquiry in geography and climate change education.

Chapter 13
Reflecting on Field-Based and Technology-Enabled Learning in Geography

Chew-Hung Chang, Kim Irvine, Bing Sheng Wu and Tricia Seow

Abstract The chapter provides a summary of the various perspectives presented in this edited book volume and provides some additional concluding thoughts. The theoretical perspectives and contemporary issues and research on the topics of teaching and learning using ICT will be reiterated and explored further in the context of NIE and Singapore. There are also several chapters in the book that describe the advantages of integrating the two approaches for teaching and learning geography. The concluding chapter will end off with a critical commentary and some recommendations on the ways that teaching and learning beyond the traditional classroom can be conducted.

Why Do Geography Teachers Take Their Lessons Beyond the Classroom?

As the authors reflected on the purpose of extending teaching and learning geography beyond the classroom, there was a consensus that geography must be lived and experienced. It has to be relevant to the lives of our students, and this cannot be done solely through textbooks and pen and paper tasks. While there are logistical concerns whenever we bring students outside the confines of the school, the potential benefits of learning and living geography far outweigh inconveniences. Nevertheless, the authors have all shared a common experience, in this regard.

C.-H. Chang (✉) · K. Irvine (✉) · T. Seow (✉)
National Institute of Education, Nanyang Technological University, Singapore, Singapore
e-mail: chewhung.chang@nie.edu.sg

K. Irvine
e-mail: kim.irvine@nie.edu.sg

T. Seow
e-mail: tricia.seow@nie.edu.sg

B. S. Wu (✉)
National Taiwan Normal University, Taipei, Taiwan
e-mail: wbs@ntnu.edu.tw

Afternoon thunderstorms are common in Singapore, where the authors come from. While we have all conducted fieldwork with wet weather plans, we have been inevitably constrained by a sudden downpour. An anecdote from one of the authors describes a typical scene. He was writing this section inside a pavilion (or was really just jotting down the notes) while waiting for the rain to stop at Changi Point Beach, Singapore. At that point, he was preoccupied with the thought that the worst thing for an educator to do was to preach what he does not practice. So there he was, completely drenched at Changi Spit, with 36 students and 3 teachers who had just completed 2 h of fieldwork asking the question why the beach needed soft engineering protection through beach nourishing. The real reason why he was there in the field was to follow through a series of professional development workshops as part of the teacher preparation for the implementation of the fieldwork component in a new syllabus for school geography in 2013. He wanted to see if the training had been well received and how teachers would take what they had learnt to put it into practice. So he contacted one teacher who attended the course and decided to follow them on the fieldwork to understand how she planned and implemented the session. The thought that whether teaching and learning geography beyond the traditional classroom is carried out well is what all geography educators are interested in.

Fieldwork in Learning Geography

Fieldwork has been considered a hallmark of geographic education by teachers and researchers alike. Much has been written about fieldwork and its place in geography. In the literature review by Kent et al. (1997) on the issue of the effectiveness and importance of fieldwork in geographic education, field studies were found to provide the integration of the theoretical with practical concepts taught in the classrooms. In addition, Kent et al. (1997) proposed that fieldwork is commonly accepted as a process that encourages holistic geographical understanding of issues. The chapter "Learning in the field—a conceptual approach" in this book has addressed the reason why fieldwork is important for geography, what successful learning of geography through fieldwork looks like and the importance of subject disciplinary thinking in geography fieldwork. The issue of subject matter knowledge versus real-world phenomenon is brought to the fore in the chapter "What happened to the textbook example of the Padang Benggali groyne field in Butterworth, Penang?" where the author urges teachers to verify the accuracy, existence and relevance of the site, lest the landforms or buildings that are intended for geographical investigation have vanished.

Moreover, some teachers commonly conduct fieldwork as field trips where they are in reality just tours or excursions (Chang and Ooi 2008). Students remain largely passive and assume the roles of tourists. Inevitably, these field trips can be described as boring, as students are not deeply engaged in the fieldwork process (Brown 1969). Yet counter-arguments suggest that properly organized fieldtrips can provide students with experiences, knowledge, understanding as well as skills that are important to an understanding of the world around them (Kent et al. 1997, p. 315). The chapter "Paradigm Shift in Humanities Learning Journey" introduces a *Balanced Scorecard* to ensure that the intended outcomes of fieldwork was to provide an immersion of the experience, vis-à-vis the need to have purposeful tasks designed to guide the process are delicately balanced.

Taking a students' perspective, all field activities can be described in two dimensions. "First, between observation and participation; second, between dependency and autonomy" (Kent et al. 1997, p. 316). Undeniably, fieldwork usually involves several different combinations of activities on these two dimensions to give students the opportunity to experience and make sense of the social, cultural and environmental phenomena (Gerber and Goh 2000). For instance, Couch (1985) suggested that meaningful learning can result from well-structured observations if it is reinforced by on-site discussion. The example from the chapter "The River Guardian Program for Junior High Schools on the "River of Kings", Thailand" shows that students can learn through meaningful activity, observe and discuss what they have found and make meaning of the geographical site. In this case, the students were able to develop a strong sense of empathy for environmental care for the river as they became deeply involved with the water quality management process in which they have taken part.

In practice, much of the fieldwork conducted by teachers falls somewhere in the middle on both dimensions. While Chew (2008) and Chang and Ooi (2008) separately argue for the affective aspect of understanding the context of a site, the later work by Chang and Seow (2010) focused on the inquiry process as a means of making fieldwork meaningful. Fieldwork is seen as a balance between observations versus participation as well as guided versus self-directed learning (dependency and autonomy in Kent et al.'s (1997) terms).

In particular, Chang and Seow (2010) proposed to adopt an inquiry approach in which four steps were outlined: i) to identify the issue and develop a question; ii) to gather and collect data; iii) to process and reorganize the data; iv) and to reflect and make sense of the information collected. While this simple approach is common to most inquiry-based learning, it has provided the framework against which fieldwork can take place in school geography. This theme is also featured in the chapter "Location-aware, context-rich field data recording, using mobile devices for field-based learning in Geography". While the chapter is about the use of technology, it is a bridging chapter to the next section of learning geography beyond the traditional classroom—and in this case, through using technology. The technology-enabled fieldwork is designed around the same cycle of inquiry proposed by Chang and Seow (2010).

Pre-service teacher education is one of the primary mandates of NIE, and the geography programme is housed within the Humanities and Social Studies Education Academic Group. There are two components of the geography programme that are noteworthy in the context of this book. First, the BA programme *requires* that the fieldwork conducted for the final year project be done overseas. Typically, the fieldwork takes place over a two-week period and past field sites have included Malaysia, Taiwan, Vietnam, Cambodia, Thailand, China, Australia and the USA. The focus of each field effort will vary between physical and human geography, but all trips include some elements of both sides of the discipline. Examples of these field efforts are discussed nicely in the chapter by Das and Chatterjea. Of course, there can be a long discussion regarding the value-added of conducting fieldwork overseas and this issue, in the context of the NIE final year project, is discussed in detail by Irvine et al. (2015). One of the authors of this chapter who take turns in offering the final year project course requires the students to keep a daily log of thoughts and observations. These logs are not meant to be the mechanical recording of the individual research tasks that were accomplished on a particular day, but rather a reflection on what each student observes in their surroundings and their reactions and feelings regarding these observations. Upon return to NIE, the logs are collected and assessed by the instructor, who also identifies common or important points to be discussed during follow-up classes. The observations can be quite insightful, details of which are discussed by Irvine et al. (2015). However, as one student eloquently noted:

> By going overseas, students/researchers/undergraduates get to observe and immerse themselves in a culture that they are not used to, making them more culturally aware. In a globalized world of today, it is important that students become culturally sensitive and globally aware of the people living in other countries. This is especially important for student teachers who would one day lead their own students for a Geographical Inquiry field trip to other countries. One of Singapore's desired outcomes of education is "be able to collaborate across cultures and be socially responsible" (Ministry of Education Singapore 2010). Another one of Singapore's desired outcomes of education at the end of post-secondary education is "be proud to be Singaporeans and understand Singapore in relation to the world" (Ministry of Education Singapore 2010). In order to better understand Singapore in relation to the world, it is necessary that students are exposed to the world first.

It also has been gratifying to see that the students frequently identify themselves as "geographers" in these reflections. In the 2014 class, for example, 39% of the students explicitly identified themselves as "geographers". It seems to us that the programme is doing something right if the students have been instilled with this sense of pride and identity in being a geographer:

> Hence, with this project, I became to realize that as a geographer, I could not leave the condition of our environment to higher authorities.

Another student noted:

> As a geographer, it is essential that our skills and understanding of geography encompasses both the local and international setting as we are able to learn and apply transferable techniques, as well as realize that not all studies conducted in one country can be applied to another.

The international experience seems to instil a sense of comradery in the field and a sense of accomplishment in stretching oneself beyond the comfort zone. Of course, the international field programme is not without its challenges. Supervising a recent cohort, one of the authors, exasperated by a general lack of resiliency in the field, posed the question at the nightly debriefing "why does NIE require you to conduct overseas fieldwork?". A number of the responses in the submitted daily logs were fairly pedestrian, "exposed to different types of physical conditions that can't be seen in Singapore", but a couple of the students described the benefit of "living the life of a geographer" (including for this student, having the runs for a couple of days!!) and "pushing outside the comfort zone to discover strengths, confidence, self-understanding." Sometimes even trying situations can translate well into teachable moments.

The second component of the geography programme that is noteworthy is its emphasis on providing authentic learning experiences in the field. Faculty in the programme have strong connections with agencies in Singapore, including the Public Utilities Board (PUB, Singapore's water agency), nParks, the Urban Redevelopment Authority, the Agri-Food and Veterinary Authority, and Jurong Town Council Corporation. These connections allow faculty to bring their own research to the classroom, but also facilitate interesting experiences in the field. For example, the PUB recently funded the construction of a large raingarden on the NIE campus. Students in the AAG33C (ecosystem dynamics) class were involved in the characterization of a candidate site as part of one of their assignments, measuring infiltration rates, soil texture, organic content, landscape slope and drainage patterns (Fig. 13.1a). The consultant designing the raingarden provided a guest lecture to discuss the design considerations for a raingarden with the class, and the students then were able to observe the construction process (Fig. 13.1b). The next year's batch of students is now beginning to conduct post-construction monitoring, measuring inflows and outflows from the garden to develop a water budget, as well as sampling water quality to determine the benefits of the raingarden. This experience gave students a rare opportunity to witness and participate in an environmental project from start to finish.

From the two dimensions of participation and observation to guided and self-directed by Kent et al (1997), to the need for contextual understanding of the site, understanding the geography of a place, and the need to frame the field experience within an inquiry approach, the chapters in this book have provided a range of examples in field-based geography teaching and learning, for the geography teacher to think about. Nevertheless, as Favier and van der Schee (2009) have suggested, students' learning can be greatly enhanced by allowing them to investigate real-world problems by combining fieldwork with ICT—GIS in their study (Favier and van der Schee 2009).

Learning Geography with ICT

Chang and Hedberg (2007) argue that how effectively students learn geography with ICT "depends largely on how [the] found resources are used, and the way the learning activity is designed" (Chang and Hedberg 2007 p. 60). Indeed, Favier and van der Schee have argued that education in the past few decades has increasingly focused on a shift from "acquisition of knowledge to the development of skills

Fig. 13.1 a (left) AAG33C students measuring infiltration rates at a candidate site for the raingarden, **b** (below) the start of construction, excavation and completion of the raingarden, with students surveying vegetation on raingarden handover

required to gain knowledge" (Favier and van der Schee 2009 p. 261) While the example was based on using information technology in fieldwork, there is much to be said about its value with or without going out to the field.

The chapter "Teaching geography with technology—a critical commentary" provided a framework to present the examples of how ICT is used in geography learning. While there was a heavy focus on the role of geographical thinking in helping teachers make decisions on what to teach, how to teach and how to assess what students have learnt, there was also an equal focus on tapping the relevant affordances of the technology. Indeed, the chapter introduced the idea of Technological Pedagogical Content Knowledge (TPACK)—the nexus between what we need to teach, who we are teaching, and how we teach. In particular, we are not just interested in using technology, but we are interested in matching the learning activity we design to student's needs and the geographical knowledge and skills that have to be taught. In the chapter "High-speed Mobile Telecommunication Technology in the Geography Classroom", the examples highlighted how technology can take the students beyond the classroom to the contexts of the expert geographer who may be at a field site, making learning authentic.

TPACK describes the relationship between the teacher, school geography and the student's learning experiences within the curriculum making process suggested by Lambert and Hopkin (2014). The geographical PCK will inform how teachers interpret syllabus documents and develop unit and lesson plans for their students. Indeed, the geographical knowledge will drive the pedagogy and consequently the way the teacher will choose which technology to use. It is important for teachers to be able to determine how they can help students think geographically so as to take them beyond what they already know and "how we have come to know it" (Lambert and Hopkin 2014, p. 65).

But there are some special areas in ICT that support geography teaching and learning best. Maps are used traditionally as a spatial representation of reality, and are often used to support students' geographical thinking. Developments in ICT have allowed maps to become a lot more portable as well as providing enhancements to this spatial representation. In addition to having mobile devices that provide real-time maps, ICT can enable a representation beyond two-dimensional spatial phenomenon and patterns. This allows students to collect, retrieve spatial information and construct spatial knowledge. The interaction afforded by technology also lifts the limits on the static representation of paper maps. There are many examples of how various geospatial technologies (GSTs) have been developed and adopted in geography education to enhance students' geographical skills and ability to think spatially (Bednarz 2004).

In addition, there is a growing number of options that children can learn geography from informal sources, such as through Internet resources and even through self-directed learning on massive open online course (MOOC). The first MOOC appeared in 2008, and they can be classified as cMOOCs and xMOOCs. The former features participation-based learning where individuals create authentic projects (usually in smaller numbers), and the latter have massive numbers of learners being enrolled per course, and usually foreground content mastery (Fasimpaur 2013). These courses are typically targeted at adult learners, while this book focuses on school geography; the authors are not ignoring the potential that MOOCs have for learners in school. At the least, it affords an additional professional learning platform for teachers who can then deepen their TPACK in designing meaningful learning of geography students in school.

Regardless of whether the ICT used for geography education is supported by the TPACK of the geography teacher, or if it taps into the spatial representation and manipulation affordances of GSTs, the common problem encountered by most teachers is that technology is often developing too quickly for them to catch up on. The idea presented earlier in this book is that instead of playing catch up, teachers can stay attuned to the latest developments not by adopting new technologies wholesale, but rather identify how they could customize the new tools for their existing needs. The chapters on using social media and authentic learning demonstrate the need for innovation in the way technology can be adapted beyond their original contexts for geographical learning.

At NIE, we have developed the Sustainability Learning Laboratory (SLL) to help Singapore geography teachers adapt to new technologies. The SLL incorporates both virtual and outdoor physical laboratories to help teachers address the concepts of sustainability presented in the Singapore geography curriculum. The virtual laboratory (www.ssl.com) provides videos, handbooks, lesson plans and other teacher resources related to the themes of water (quantity and quality), climate and climate change, urban liveability, tourism, transportation and food security. The virtual laboratory also provides a geospatial data portal that includes simple Web-based mapping and digital data layers such as land use and remotely sensed surface temperatures in Singapore, as well as data storage, retrieval and visualization of environmental time series data monitored in the outdoor physical laboratories. Currently, the outdoor physical laboratories include the NIE raingarden and the Jurong Eco-garden. Data monitoring includes meteorological stations that record rainfall, wind speed, wind direction, atmospheric pressure, relative humidity and shortwave radiation; air quality monitoring for sulphur dioxide, nitrogen dioxide, carbon dioxide, carbon monoxide and particulates (2.5/10 pm), as well as noise, and water quality using YSI datasondes to measure temperature, dissolved oxygen, conductivity, pH, turbidity and fluorescence (chlorophyll *a*).

The outdoor physical laboratories also provide opportunities to conduct fieldwork, in keeping with the themes of the previous sections of this chapter. For example, the Jurong Eco-garden includes a series of storm water detention ponds and a cleansing biotope to manage the storm water following water-sensitive urban design (WSUD) principles. NIE students, as well as students from nearby partner schools (primarily Secondary 1 level), have been exploring the efficacy of these ponds by routinely testing water quality parameters that include dissolved oxygen, temperature, turbidity, nitrates, phosphates, E. coli and pH (Fig. 13.2). The sites at which the sampling is done can be georeferenced, and a Water Quality Index is calculated using the free app, WaterScope, available from the app store. This app was developed by Mr. Vernon Tan for his final year project in geography at NIE. The georeferencing for the sample sites is similar in principle to that described in the chapter "Authentic Learning: making sense of the real environment using mobile technology tool", although WaterScope specifically focuses on water quality. The SLL Website explains both the use of the WaterScope app and the theory behind a Water Quality Index.

The concept of continuously monitored data at the outdoor physical laboratories is consistent with society's growing interest in "big data" and the implications for smart cities. "Big data" refers to a massive volume of both structured and unstructured data that is so large that it is difficult to process using traditional database approaches. Sensors collecting air and water quality data every 15 min at multiple sites, year in and year out, would constitute a component of "big data", although some would argue this is just a new buzzword for the same old thing we have been doing for decades—automated monitoring. Chen et al. (2014) note that big data can be used in many applications, including enterprise management, marketing, the Internet of things, and smart grids (including water, power and transportation grids). A number of the sensors connected to the SLL Website also are connected to the Internet of things. Many big data sets have inherent geospatial structure, and just as geographers actively pursued the quantitative revolution in the 1960s, geographers today should embrace the analytical needs of big data. This underscores the importance of quantitative skills and means that geography teachers will need to become comfortable with applied numerical methods that can be used in their classrooms.

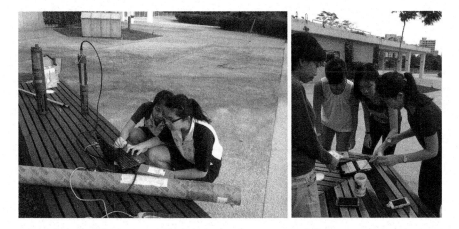

Fig. 13.2 River Valley JC1 geography students (left) calibrating YSIs at the Jurong Eco-garden and NIE BA geography students (right) analysing water samples from Jurong Eco-garden ponds for nutrients and E. coli

In considering the closing remarks to the chapter "Teaching geography with technology—a critical commentary", we cannot avoid the fact that our children are very savvy with technology, cannot and will not be able to get through a day in their lives without a mobile device. Consequently, there is no escaping the use of ICT in our children's learning experiences, as it is an integrative part of our students' reality. Rather than debating whether it is possible to learn geography with or without technology, it is imperative that geography educators should examine how ICT can be used meaningfully, instead.

Conclusion

Geography is concerned with "the study of Earth and its natural and human environments. Geography enables the study of human activities and their interrelationships and interactions with environments from local to global scales" (International Geographical Union—Commission on Geographical Education 2016, p. 4). Teaching and learning about geography within and beyond the traditional classroom should be about understanding these concepts, for the purposes of educating our children for a better world. The issues of assessing learning through ICT or fieldwork have been deliberated in the chapter "Assessing how Geography is learnt beyond the classroom". Besides designing assessment tasks and items that are reliable and valid, the authors argued for assessment with consequential validity. Indeed, if geography is about understanding nature and humans, the interrelationships and interaction across scales, then it would be inevitable that the learning should take place beyond the classroom into the field, and even to venture into "places" beyond, maybe vicariously through technology.

As the author was collecting his thoughts on the notes first outlined in this chapter, still drenched at Changi Point Beach he was heartened by questions asked by the students. They wanted to know "How do we know that the boats did not affect our readings?" or "But the uncle just swept the sand and we don't know where to start our profile". While it could be argued that the professional development programme might or might not have helped the teacher conduct the fieldwork well, the observations point to how actual teaching practice is linked to the intentions of geography education. In referring back to Michael Young's idea of a F3 curriculum, the authors cannot help but reflect on how geographical knowledge would allow the child to engage new information critically, ask questions about the information based on geographical knowledge and develop new ways of thinking, analysing, explaining and understanding, as well as take control of his/her own knowledge and take part in international debates on issues, thereby enabling him/her to engage key issues in society. In other words, it is not about teaching or learning about geography, but rather educating a child for the world. The chapters collected in this book volume have spanned across topics and countries in Southeast Asia, providing examples for geography educators to reflect on how their research and practice can be advanced. However, the unifying theme that runs through all the chapters is that geography matters, for geography enables a child to develop capabilities to do well in the world that they live in. Consequently, the authors hope that whether geography is taught and learnt within or beyond the traditional classroom, it must make a difference to our children's lives in a volatile, uncertain, complex and ambiguous world.

References

Bednarz, S. W. (2004). Geographic information systems: A tool to support geography and environmental education? *GeoJournal, 60*(2), 191–199.
Brown, E. H. (1969). The teaching of fieldwork and the integration of physical geography. In R. U. Cooke & J. H. Johnson (Eds.), *Trends in geography: An introductory survey* (pp. 70–78). London: Heinemann.
Chang, C. H., & Hedberg, J. G. (2007). Digital libraries creating environmental identity through solving geographical problems. *International Research in Geographical & Environmental Education, 16*(1), 58–72.
Chang, C. H., & Ooi, G. L. (2008). Role of fieldwork in humanities and social studies education. In O.S. Tan, D.M. McInerney, A.D. Liem, & A.G. Tan (Eds.), *Research in multicultural education and international perspectives series, Volume 7: What the West can learn from the East. Asian perspectives on the psychology of learning and motivation* (pp. 295–312). Charlotte: Information Age Publishing. ISBN: 9781593119874.
Chang, C. H., & Seow, T. D. I. C. (2010). Field inquiry for Singapore geography teachers. In C.H. Chang, L.C. Ho, T.D.I.C. Seow, K. Chatterjea, & G.C.I. Tan (Eds.), *Understanding the changing space, place and cultures of Asia—SEAGA 2010 online proceedings*. Singapore: Southeast Asian Geography Association.
Chen, M., Mao, S., & Liu, Y. (2014). Big data: A survey. *Mobile Networks and Applications, 19*(2), 171–209.
Chew, E. (2008). Views, values and perceptions in geographical fieldwork in Singapore schools. *International Research in Geographical and Environmental Education, 17*(4), 309–329.
Couch, I. R. (1985). Fieldwork skills—The potential of foreign environments. In R. Barass, D. Blair, P. Garnham, & A. Moscardini (Eds.), *Environmental science teaching and practice, Conference Proceedings: Third Conference on the Nature and Teaching of Environmental Studies and Sciences in Higher Education* (pp. 247–252). Northallerton, England: Emjoc Press.
Fasimpaur, K. (2013). *Massive and open* (pp. 12–17). March/April: Learning & Leading with Technology.
Favier, T., & Van Der Schee, J. (2009). Learning geography by combining fieldwork with GIS. *International Research in Geographical and Environmental Education, 18*(4), 261–274.
Gerber, R., & Goh, K. C. (Eds.). (2000). *Fieldwork in geography: Reflections, perspectives and actions*. Dordrecht, The Netherlands: Kluwer Academic.
International Geographic Union—Comission on Geographical Education. (2016, August). *International charter on geographical education*. Retrieved August 2016, from International Geographic Union—Comission on Geographical Education: http://www.igu-cge.org/Charters-pdf/2016/IGU_2016_def.pdf.
Irvine, K. N., Seow, T., Leong, Ka Wai, & Cheong, S. I. D. (2015). How high's the water, mama? A reflection on water resource education in Singapore. *HSSE Online, 4*(2), 128–162.

Kent, M. Gilbertson D. O., & Hunt, C. O. (1997). Fieldwork in geography teaching: A critical review. *Journal of Geography in Higher Education, 21*(3), 313–332.

Lambert, D., & Hopkin, J. (2014). A possibilist analysis of the geography national curriculum in England. *International Research in Geographical and Environmental Education, 23*(1), 64–78.

Chew-Hung Chang is a geography educator serving as the co-chair of the International Geographical Union, Commission on Geographical Education, co-editor of the journal International Research in Geographical and Environmental Education, as well as the President of the Southeast Asian Geography Association. In addition to being a teacher educator, Chew-Hung has published extensively across areas in geography, climate change education, environmental and sustainability education.

Kim Irvine has worked in the field of hydrology and water resources for more than 30 years. He was awarded the New York Water Environment Association, Environmental Science Award, in 2013. His work has ranged from detailed regulatory studies to capacity building workshops on water quality sampling and assessment for universities, NGOs and government agencies throughout North America and Southeast Asia. His research interests include urban hydrology, water resource management, water quality and applied urban drainage modelling.

Bing Sheng Wu has an expertise in Geographic Information Science and Geospatial Technologies. He teaches across undergraduate to graduate-level courses at the National Taiwan Normal University and has been actively researching on applied aspects of Geospatial technologies for humanities education.

Tricia Seow As an experienced geography educator, Tricia Seow is involved in the iGeog Steering Committee and Talent Development Programme Fieldwork Trainer, Southeast Asia Geography Association, and the External International Consultant for MA Education Programme. She enjoys fieldwork, and her research interest includes teachers' knowledge and practice, field inquiry in geography and climate change education.

Index

A
Assessment, 3, 8, 43, 49, 50, 52–57, 104, 119, 126, 148, 210
Authentic learning, 8, 38, 111–116, 129, 146, 153, 154, 205, 209

B
Balanced Scorecard, 101, 102, 104, 105, 203

C
Chao Phraya River, 80, 82–85, 87–89, 93
Classroom, 3, 8, 9, 42, 49–52, 54–57, 63, 83, 102, 104, 107, 112, 113, 116, 124–127, 129, 145–149, 151–154, 163, 164, 166, 170, 201, 203, 207, 210, 211
Comfort level, 173–175, 177, 182, 183, 188, 193–195
Conceptual approach, 11, 35, 202
Consequential validity, 50, 52, 54, 55, 57, 210
Critical thinking, 12
Curriculum making, 8, 42, 43, 51, 52, 57, 207

D
Data, 7, 11, 15–21, 23, 27, 30, 39–42, 45, 54–56, 84, 90, 93, 104, 111, 112, 114–122, 124–127, 129, 133, 135–137, 139–142, 149, 152, 154, 165–168, 170, 180, 182, 183, 187, 190, 203, 208, 209
Data analysis, 115, 126, 127, 141

E
Education, 4–6, 9, 36, 39, 45, 49, 51, 56, 57, 82, 83, 85, 95, 96, 102, 106, 107, 113, 147, 153, 202, 205, 207, 208, 210
Education for sustainable development, 85

Examinations, 49, 55, 57, 146, 147

F
Field-based, 3, 6, 18–24, 26, 50, 55, 57, 96, 101, 111–115, 118, 125, 127–129, 136, 137, 203
Fieldwork, 3, 4, 6–9, 45, 50, 57, 63, 65, 83, 85, 111, 113, 114, 116–119, 124, 128, 137, 164, 202, 203, 205, 207, 210

G
Geographical education, 4, 56
Geography, 3–9, 35–45, 49, 50, 52–57, 63, 96, 101, 128, 129, 134, 136, 142, 145–147, 163, 164, 169, 170, 201–203, 205, 207, 208, 210
Geography teachers, 5, 30, 173, 175, 176, 181–184, 188–195, 208, 209
Groyne field, 64–66, 68, 69, 73–76, 202

H
Humanities, 101, 145–148, 150–152, 165, 175, 180, 182, 191–193, 195, 203

I
Information, 5–7, 35–42, 45, 50, 54, 55, 69, 90, 112, 114–117, 120, 122, 124, 133, 134, 136–138, 140–142, 148, 156, 163, 164, 166, 168, 169, 203, 207, 210
Information and Communication Technology (ICT), 7–9, 35, 38, 39, 43–45, 49, 50, 53, 141, 163, 164, 168, 169, 173–179, 181–183, 186–196, 201, 205, 207, 208, 210

Inquiry, 6, 36, 55, 133, 136, 137, 139–141, 149, 152, 163–170, 203, 205

K
Knowledge, 4–8, 35–40, 42–44, 49, 52, 53, 55–57, 85–87, 93, 112–117, 122, 124, 126, 128, 129, 135, 139, 142, 145–147, 155, 156, 163, 168–170, 202, 203, 205, 207, 210

L
Learning, 3–9, 35–40, 42–45, 49, 50, 52–57, 63, 65, 68, 82–85, 87, 95, 96, 101–107, 111–118, 126–128, 133–137, 139–141, 145, 146, 148–154, 163, 164, 168–171, 195, 201–203, 205, 207, 208, 210
Learning Journey, 102–106, 146, 203
Location-aware, 115–117, 124, 141, 203

M
Mobile, 7, 8, 39–42, 56, 111–116, 118–121, 125, 129, 133, 134, 136, 137, 140, 142, 145–155, 203, 207, 210
Mobile devices, 40, 115, 133, 136, 142, 146, 148

P
Physical geography, 19

R
Research, 4–6, 8, 40, 76, 82, 93, 96, 111, 114, 115, 117, 120, 121, 125, 126, 129, 134, 137, 140, 147, 149, 168, 169, 201, 210
River Guardian program, 80, 82

S
Schools, 5, 14, 79, 82–84, 86–88, 91, 96, 101, 102, 105, 106, 113, 114, 127, 133, 137, 149, 174–176, 178, 181–183, 192, 193, 203, 208
Social media, 35, 41, 163–166
Students, 4–8, 11–30, 35, 36, 38–40, 42–45, 49–58, 63–66, 79, 82, 83, 85–88, 90–93, 95, 101–107, 112–116, 118–120, 122, 124–129, 133–137, 139–142, 145–151, 153–156, 158, 163–171, 173, 174, 176, 178–189, 191, 192, 195, 201–205, 207, 208, 210

T
Teachers, 3, 5–8, 11, 13, 14, 17–19, 29, 30, 37, 38, 42–44, 49–53, 56, 57, 63–65, 76, 79, 83, 87, 91, 93, 95, 96, 101–103, 105–107, 114, 126–128, 134, 137, 139–142, 145–150, 152–154, 173–182, 184, 187–192, 194–196, 202–204, 207, 208
Technology, 3, 4, 6, 7, 35–39, 41–45, 56, 111–113, 115, 116, 129, 134, 136, 137, 139–142, 145, 147–151, 153, 154, 195, 203, 207, 208, 210
Textbook, 63–65, 76, 148, 151, 168, 169, 202
Tiong Sa Teh, 63

W
Water quality, 79, 82–84, 86, 87, 89–91, 203, 205, 208, 209
Wordpress, 163